W0254299

Climate Change, Soil and Agricultural Technologies for Sustainable Development, Food and Social Security

Climate Change, Soil and Agricultural Technologies for Sustainable Development, Food and Social Security

Subhash Chand
Assistant Professor
Division of Soil Sciences
Sher-e-Kashmir University of Agricultural Sciences and Technology of Kashmir, Shalimar Campus
Srinagar 191121
Jammu and Kashmir

CBS Publishers & Distributors Pvt Ltd

New Delhi • Bengaluru • Chennai • Kochi • Mumbai • Pune
Hyderabad • Kolkata • Nagpur • Patna • Vijayawada

Disclaimer

Science and technology are constantly changing fields. New research and experience broaden the scope of information and knowledge. The author has tried his best in giving information available to him while preparing the material for this book. Although, all efforts have been made to ensure optimum accuracy of the material, yet it is quite possible some errors might have been left uncorrected. The publisher, the printer and the author will not be held responsible for any inadvertent errors, omissions or inaccuracies.

Climate Change, Soil and Agricultural Technologies for Sustainable Development, Food and Social Security

ISBN: 978-81-239-2494-6

Copyright © Author and Publisher

First Edition: 2015

All rights reserved. No part of this book may be reproduced or transmitted in any form or by any means, electronic or mechanical, including photocopying, recording, or any information storage and retrieval system without permission, in writing, from the author and the publisher.

Published by Satish Kumar Jain for
CBS Publishers & Distributors Pvt Ltd
4819/XI Prahlad Street, 24 Ansari Road, Daryaganj, New Delhi 110 002, India.
Ph: 23289259, 23266861, 23266867 Website: www.cbspd.com
Fax: 011-23243014 e-mail: delhi@cbspd.com; cbspubs@airtelmail.in.
Corporate Office: 204 FIE, Industrial Area, Patparganj, Delhi 110 092
Ph: 4934 4934 Fax: 4934 4935 e-mail: publishing@cbspd.com; publicity@cbspd.com

Branches

- **Bengaluru:** Seema House 2975, 17th Cross, K.R. Road, Banasankari 2nd Stage, Bengaluru 560 070, Karnataka
  Ph: +91-80-26771678/79 Fax: +91-80-26771680 e-mail: bangalore@cbspd.com
- **Chennai:** 20, West Park Road, Shenoy Nagar, Chennai 600 030, Tamil Nadu
  Ph: +91-44-26260666, 26208620 Fax: +91-44-42032115 e-mail: chennai@cbspd.com
- **Kochi:** 36/14 Kalluvilakam, Lissie Hospital Road, Kochi 682 018, Kerala
  Ph: +91-484-4059061-65 Fax: +91-484-4059065 e-mail: kochi@cbspd.com
- **Mumbai:** 83-C, Dr E Moses Road, Worli, Mumbai-400018, Maharashtra
  Ph: +91-22-24902340/41 Fax: +91-22-24902342 e-mail: mumbai@cbspd.com
- **Pune:** Bhuruk Prestige, Sr. No. 52/12/2+1+3/2 Narhe, Haveli (Near Katraj-Dehu Road Bypass), Pune 411 041, Maharashtra
  Ph: +91-20-64704058/59, 32392277 Fax: +91-20-24300160 e-mail: pune@cbspd.com

Representatives

- **Hyderabad** 0-9885175004
- **Nagpur** 0-9021734563
- **Vijayawada** 0-9000660880
- **Kolkata** 0-9831437309, 0-9051152362
- **Patna** 0-9334159340

Printed at Paras Offset Pvt. Ltd, New Delhi

To

My beloved parents
Sh Teekam Singh
and
Smt Maya Devi

Preface

Sustainable development of world agricultural food and social security is of prime concern. World debate on decreasing factor productivity, climate change, global warming, quality and productivity of major crops and cropping systems, economic crunch, strategic society development, poverty and hunger is intensified. Sustainable development means the development of agriculture in a manner of eco-friendly, ecologically balanced sociologically adaptable and economically acceptable by farmers and users for maintaining good soil health, and ecological harmony of flora and faunaa. For meeting challenge of food and social security and sustainable development, several modern agricultural technologies like precision farming (PF), integrated nutrient management (INM), resource conserving technologies (RCTs), integrated pest management (IPM), organic farming (OF), value addition, site specific nutrient management (SSNM), soil carbon sequestration (SCS) widely researched and adapted all over the world.

Soil suitable for cultivation is precious and a finite resource. Irreversible degradation of this resource implies not only ruining the main assest of current farmers, but also reducing the farming opportunities of future generations. Therefore soil protection policies need to have a special focus on sustainable use and management of agricultural soils with a view to safeguarding the fertility and agronomic values of agricultural land. Sustainable and equitable development without further degradation in natural resources are of prime concern.

The book "*Climate Change, Soil and Agricultural Technologies for Sustainable Development, Food and Social Security*" is a compilation of best soil, nutrient, crop and climate change mitigation *vis-à-vis* management strategies and carbon sequestration, resource conservation agriculture practices in changing world agriculture scenario due to pressure on land and water for sustainable food security and socio-economic development of evergrowing world population. This book has 10 chapters on various issues related to modern soil and crop science coupled with sustainable agriculture resource conserving technologies. Each chapter is well described and written consistently to clear the basic and strategic concept with suitable examples, important resources and references in the last for further reading.

Chapter 1 deals with global climate change, causes of climate change, impact of climate change on agriculture and systems, climate change and water, mitigation and adaptation strategies for combating global climate change including land uses practices (LUPs), integrated soil fertility management (ISFM), biological management of soil fertility (BMSF), role of biodiversity and resource conservation technologies (RCTs).

Chapter 2 highlights the soil health (quality) *vis-à-vis* soil organic carbon in scenario of climate change and challenges of food security. The enhancing soil organic carbon stock (SOC) looks an ultimate solution for sustainable soil health and food security.

Chapter 3 focuses on soil carbon sequestration as solution for mitigating global climate change along with soil carbon sequestration practices like adoption of recommended agricultural practices (RAPs) and land use practices (LUPs) for uses and benefits of research scholar, soil scientist, crop scientist, agronomist, extension worker and farm workers.

Chapter 4 describes carbon farming and sequestration for enhancing soil organic carbon (SOC) in agricultural lands for sustainable crop production and productivity. Various carbon farming practices like low tillage/no tillage, crop rotations, fertilizations, land use management (LUM), biochar burial and composting technologies has been described with suitable examples.

Chapter 5 is a classical collection and compilation of A to Z information on resource conservation technologies (RCTs) for sustainable agriculture. It gives all the answers related to conservation agriculture in modern changing agriculture scenario like frequently ask questions mode, e.g. goal of CA, characteristics of CA, hat is not CA, is CA compatible with IPM, role of animal husbandary in CA, down sides of CA, benefits of CA, issues in CA, is conservation agriculture real, new machines in CA, RCTs potential tools for attaining food, nutritional and livelihood security, RCTs in rice-wheat system and need for CA, laser land leveling, RCTs in wheat, zero tillage and energy and economics of CA.

Chapter 6 deals with approaches in fertiliser recommendations for maximising yield and sustainable soil health with highlighting soil testing services in India, advantage of soil testing, approaches in formulations of fertiliser recommendations like generalised recommendations, fertiliser recommendations based on soil fertility categories, soil test based fertiliser recommendations for a certain percentage of yield, fertiliser recommendations based on soil critical limits, soil test based recommendations (STCR) for target yield of crops, target yield concept and adjustment equation, STCR under IPNS for optimising doses, IPNS vs fertiliser alone on crop response, site specific nutrient management (SSNM) for rice crop, diagnosis and recommendations integrated system (DRIS) for the uses of soil and crop experts *vis-à-vis* progressive farmers.

Chapter 7 well describes concept, objectives, components, and limitations of INM with classical examples. This chapter suggest appropriate and best INM options for important crops viz., rice, wheat, sorghum, maize, pearl millet, soyabean, groundnut, sunflower, cotton, mustard, sugarcane, pulses, vegetables, spices, fruit crops and ornamental plants besides important cropping system in different agro-climatic zones for sustaining their productivity on one hand and maintaining soil quality for future generation on the other hand.

Chapter 8 describes about opportunities for precision agriculture and remote sensing for sustainable agriculture highlighting variability in field, components of precision farming, techniques for identification of GIS problems, success of GIS, limitations of GIS, applications of GIS, global positioning system (GPS), system of GPS, operations, remote sensing (RS), precision land leveling, etc.

Chapter 9 is a collection of more than 100 reviews on the effect of deforestation on flora and soil properties like soil reaction (pH), organic carbon (OC) and available nitrogen, phosphorus and potassium.

Chapter 10 is miscellaneous topics of interest (MTI) focuses on resource management technologies (RMT) for rice-wheat cropping system, beneficial effect of SRI, salinity in vertisols, diversified farming technologies for lesser Himalayas, precision farming, organic farming and certification, maximising fertiliser use efficiency, soil pollution remedial measures are useful for readers, scientist, students and farmers. Important informations provided through appendices are useful for readers.

The book is extremely important for soil scientists, agronomist, climate change scientists, agrometrologists, extensionists, environmentalists, plant scientists, soil microbiologists, economists, research institutes, colleges, universities, research scholars, students, progressive farmers, plant pathologists, entomologists, social scientists and ecologists.

Subhash Chand

Acknowledgments

The inspiration, help and assistantce received from following deserves a mention as it helped the athour to crystallize his knowledge and experience while writing manuscript of book.

- Prof. Tej Pratap, H'ble, Vice-chancellor, SKUAST-Kashmir, Shalimar, Srinagar-191121, who inspired me to give best to the world agricultural and ecologcal community.
- Prof. Shafiq A Wani, Director of Research, SKUAST-Kashmir, Shalimar, Srinagar-191121. One of gental person and able researcher who inspired me for better service.
- Prof LL Somani, Former, Director Resident Instructions and Professor. My advisor in docroal studies, under whom I did my doctoral research at Rajasthan College of Agriculture, Udaipur.
- Dr SK Bansal, Director, Potash Research Institute of India, Gurgaon. Who helps me in potassium research activities and encouragement.
- Dr A Subba Rao, under whom I did research fellowship at Indian Institute of Soil Science, Bhopal where my interest on STCR was developed.
- Dr SL Mehta, Ex National Director, NATP, DDG (EDUCATION) and Vice-chancellor, MPUAT, Udaipur for encouragement during research associateship at PIU-NATP-KAB-II,New Delhi.
- Dr JP Mittal, Former, National Coordinator, TOE mode of National Agricultural Technology Project (NATP)-KAB-II, new Delhi-110112. Who inspired me to do best things with light moments.
- Dr RL Shyampura, Head, NBSS & LUP, Regional Center, Udaipur, Who extended his all helps during my post graduate degree and moral support during doctoral studies.
- Dr PC Kanthaliya, Professor, Rajasthan Collge of Agriculture, Udaipur, under whom, I learn about long term fertility experiments.
- Dr Sunil Pabbi, Principle Scientist, CCUBGA, IARI, New Delhi.
- Dr Badrul Hasan, Director, Resident Instrvctions Professor Agronomy, SKUAST-Kashmir, Srinagar.
- Dr FA Khan, Associate Professor, Plant Physiology, SKUAST-Kashmir, Srinagar.
- Dr Lal Singh, Assistant Professor, Agronomy, SKUAST-Kashmir, Srinagar.
- Dr FA Banday, Professor & HOD, Fruit Science.
- Dr Amarjeet Singh, Assistant Professor of Social Forestry.
- Dr Parmeet Singh, Assistant Professor of Crop Science.

- Dr Mustaq A Wani, Associate Professor of Soil Science.
- Dr GR Najar, Assoc. Prof. & HOD, Soils FOA, Wadoora.
- Dr SS Pathania, Assistant Professor of Entomology.
- Dr Bhisam Pal of IPL.

I acknowledge all my well wishers, scientists particularly those, who lended positive and healthy discussion during meetings conference/symposium to enrich my knowledege of science. My school teacher, advisor and guide who prove to me a torch bearer. Some people provide their paper assigmment and other resources to alleviate this compilation for a greater cause.

I am thankful to all my students for positive and fruitful feedbacks during classes, talks and farmers for their curiosity to know about the latest happenings.

Subhash Chand

Contents

Preface v

1. Global Climate Change Impact and Mitigation Strategies 1–6

1.1 Causes of Climate Change
1.2 Impact of Climate Change on Agriculture and Systems
1.3 Climate Change and Water
1.4 Mitigation and Adaptation Strategies for Combating Global Climate Change
1.4.1 Land Use Practices (LUPs)
1.4.2 Integrated Soil Fertility Management (ISFM)
1.4.3 Biological Management of Soil Fertility (BMSF)
1.4.4 Role of Biodiversity
1.4.5 Resource Conservation Technologies (RCTs)
1.5 Key References and Resources for Further Reading

2. Soil Health (Quality) *vis-à-vis* Soil Organic Carbon and Food Security for Social Development 7–13

2.1 Challenges of Soil Quality
2.2 Challenges of Food and Social Security
2.3 Enhancing Soil Organic Carbon (SOC) Stock—An Ultimate Solution
2.4 Key References and Resources for Further Reading

3. Carbon Sequestration—A Solution for Mitigating Global Climate Change/Global Warming 14–26

3.1 Soil Carbon Sequestration (SCS)
3.2 Adoption of Recommended Agricultural Practices (RAPs)
3.3 Land Use Practices (LUPs)
3.4 Conclusions
3.5 Future Line of Work
3.6 Key References and Resources for Further Reading

4. Farming Carbon and Sequestration for Enhancing Soil Organic Carbon (SOC) in Agricultural Lands and Cultivable Lands 27–41

4.1 Preamble
4.2 Farming Carbon (FC)
4.2.1 Low/No Tillage/Minimum Tillage/Zero Tillage
4.2.2 Crop Rotations/Cropping Sequence
4.2.3 Plant Nutrient Applications
4.2.4 Land Use Management Practices (LUMPs)

4.2.5 Biochar Burial
4.2.6 Composting Technologies
4.3 Conclusions
4.4 Key References and Resources for Further Reading

5. Resource Conservation Technologies (RCTs) for Sustainable Agriculture and Resource Conservation 42–60

5.1 Conservation Agriculture (CA) Relevance
5.1.1 Goal of CA/Long-term Objectives
5.1.2 Peculiarities of CA
5.1.3 Everything is not CA
5.1.4 Conservation Tillage Practices (CTPs)
5.1.4.1 Direct Palnting/Seeding/Sowing
5.1.4.2 Organic Farming in Relation to CA
5.1.4.3 Is CA Compatible with IPM?
5.1.4.4 What Is the Role of Animal Husbandary in CA?
5.1.4.5 What Are the Downsides of CA?
5.1.4.6 Benefits of CA
5.1.4.7 Carbon Sequestration (Greenhouse effect)
5.2 Issues in CA
5.2.1 Necessary Technologies are Often Unavailable
5.2.2 Is Conservation Agriculture Real?
5.3 New Machines for CA
5.3.1 Double Disc Coulters
5.3.2 Punch Planter/Star Wheel
5.3.3 Happy Seeders
5.3.4 Rotary Disc Drill (RDD)
5.4 RTCs in Reclaimed Alkali Soils
5.4.1 Immediate Objectives
5.4.2 Long-term Objectives/Goals
5.5 RCTs—Potential Tools for Attaining Food, Nutritional and Livelihood Security
5.6 RCTs in Rice-Wheat System and Need for CA
5.7 Laser Land Levelling
5.8 RCTs in Wheat
5.8.1 Zero Tillage
5.8.2 Energy and Economics
5.8.3 Reduced/Minimum Tillage
5.8.4 Rotary Tillage
5.8.5 Bed Planting
5.8.6 Surface Seeding
5.8.7 Effect of Tillage Options in *Phalaris*
5.8.8 Water Use and Savings Under Various Tillage Options
5.8.9 Eco-friendly Tillage Options (EFTOs)
5.9 RCTs in Rice for Water Saving
5.9.1 Direct Wet-seeded Rice
5.9.2 Weed Management in Direct Seeded Rice
5.9.3 Leaf Color Chart (LCC)

5.10 Conclusions for Adoption of Resource Conserving Technologies
5.11 Key References and Resources for Further Reading

6. Approaches in Fertilizer Recommendations for Maximizing Yield and Dynamic Soil Health 61–76

6.1 Soil Testing Services in India
6.2 Advantages of Soil Testing in Relevance to Agriculture
6.3 Approaches in Formulation of Fertilizer Recommendations
6.3.1 Generalized Recommendations
6.3.2 Fertilizer Reccommendations Based on Soil Fertility Categories
6.3.3 STCR for a Certain Percentage of Yields
6.3.4 Fertilizer Recommendations Based on Soil Critical Limits
6.3.4.1 Critical Limits
6.3.5 STCR for Target Yield of Crops
6.3.5.1 STCR for Correlation Approach
6.3.5.2 Objectives
6.3.5.3 Uses of STCR
6.3.6 Target Yield Concept and Adjustment Equation
6.3.6.1 Target Yield Equations Are Suitable under the Following Situations
6.3.6.2 STCR Studies under IPNS for Optimising Fertiliser Doses (OFDs)
6.3.6.3 Effect of IPNS vs Fertilizer Alone on Crop Response
6.3.6.4 Use of STCR for Suitability of Cropping System in Various Soil Orders
6.3.6.5 Site Specific Nutrient Management (SSNM) for Rice Crop
6.3.6.5.1 Establish an Attainable Yield Target
6.3.6.5.2 Effectively Use Existing Nutreints
6.3.6.5.3 Fertiliser Application to Fill the Defecit between Crop Needs and Supply Indigenous Supply of Plant Nutrients
6.3.7 Diagnosis and Recommendations Integrated System (DRIS)
6.4 Position and Future Line of Work
6.5 Conclusions
6.6 Key References and Resources for Further Reading

7. Integrated Nutreint Management (INM) Options for Sustainable Agriculture and Social Security 77–121

7.1 Concept of INM
7.1.1 Shorter Objectives of INM
7.1.2 Broader Objectives/Goal of INM
7.2 Components of INM
7.2.1 Soil Reserves (SRs)
7.2.2 Use of Mineral Fertilizers
7.2.3 Organic Sources of Plant Nutreints
7.2.3.1 Farm Yard Manure (FYM)
7.2.3.2 Composting
7.2.3.3 Crop Residues Management (CRM)

7.2.3.4 Green Manuring (GM)
7.2.3.5 Biogas Slurries (BGs)
7.2.3.6 Industrial Waste Materials (IWMs)
7.2.3.7 City Refuse (Garbage, Sewage Sludge)
7.3.3.8 Enriched City Compost
7.2.4 Biofertilizers— Sources of Plant Nutrition
7.2.4.1 Rhizobium Inoculants
7.2.4.2 Biofertilizers for Flooded Rice
7.2.4.3 Phosphate Solubilizing Microorganisms (PSMs)
7.2.4.4 Constraints to Use Biofertilizers
7.2.4.4.1 Production Constraints
7.2.4.4.2 Marketing Constraints
7.2.4.5 Precautions to Use Biofertilizers
7.2.5 Legumes in Cropping Systems
7.2.5.1 Dual Purpose Legumes (DPLs)
7.2.5.2 Legumes Intercropping (LI)
7.2.5.3 Legume in Rotations
7.3 Constraints of INM
7.4 Developmental Issues
7.5 Researchable Issues
7.6 Future Plan of Work
7.7 INM Options for Important Crops
7.7.1 Rice
7.7.2 Wheat
7.7.3 Sorghum
7.7.4 Maize
7.7.5 Pearl Millet (Bajra)
7.7.6 Soyabean
7.7.7 Groundnut
7.7.8 Sunflower
7.7.9 Cotton
7.7.10 Mustard
7.7.11 Sugarcane
7.7.12 Pulses
7.7.13 Vegetables
7.7.14 Spices
7.7.15 Fruits
7.7.16 Ornamental Plants
7.7.17 Miscellaneous Crops
7.8 The INM Options for Important Cropping Systems (Table 7.3)
7.8.1 Rice-Wheat
7.8.2 Rice-Rice
7.8.3 & 4 Maize-Wheat and Soyabean-Wheat
7.8.5 Miscellaneous Cropping Systems
7.9 Conclusions
7.10 Key References and Resources for Further Reading

8. Opportunities for Precision Farming (PF) and Remote Sensing (RS) for Sustainable Soil and Crop Management 122–145

8.1 Preamble
8.2 Precision Farming
8.2.1 Precision Farming— Why?
8.2.2 Variabilities in Field
8.2.3 Components of Precision Farming
8.2.4 Strategies for Implementation
8.2.5 Techniques for Identification of GIS Problem
8.2.6 Problems for Implementation
8.2.7 Success of GIS
8.2.8 Limitations of GIS
8.2.9 Miscellaneous Uses
8.2.10 Applications of GIS
8.3 Global Positioning System (GPS)
8.3.1 Systems of GPS (Global Navigation Satellite System)
8.3.2 GPS Operations
8.3.3 How GPS Works?
8.3.4 Cost Estimation of DGPS Network Infrastructure Development in India
8.4 Remote Sensing (RS)
8.4.1 Types of Remote Sensing
8.4.2 Stages of Remote Sensing/Principles
8.5 Precision Land Leveling
8.6 Variable Rate Technology
8.7 Site-Specific Planting
8.8 Site-Specific Nutreint Management (SSNM)
8.9 Precision Water Management (PWM)
8.10 Site-Specific Weed Management (SSWM)
8.11 Precision Insect, Pest and Diseases Management
8.12 Practical Problems in Indian Agriculture
8.13 Methodology to be Adopted
8.14 Economic Feasibility of Precision Farming in India on Agricultural Condition
8.15 Indians' Initiation and Conclusions
8.16 Key References and Resources for Further Reading

9. Impact of Deforestation on Flora and Soil Attributes—A Critical Review 146–160

9.1 Introduction
9.2 Impact of Deforestation on Flora
9.3 Impact of Deforestation on Soil Properties
9.3.1 Soil Reaction
9.3.2 Organic Carbon
9.3.3 Available Nitrogen
9.3.4 Available Phosphorus
9.3.5 Available Potassium
9.4 Conclusions
9.5 Key References and Resources for Further Reading

10. Miscellaneous Topics of Interest (MTI) 161–183

10.1 Resource Management Technologies for Rice-Wheat Production System
10.2 Beneficial Effects of Systems of Rice Intensification (SRI) Technology
10.3 Salinity in Vertisols
10.4 Diversified Farming Technologies for Rural Development in Lesser Himalayas
10.5 Importance of Precision Farming
10.6 Organic Farming and Certification
10.7 Guidelines for Maximizing Fertilizer Use Efficiency
10.8 Soil Pollution through Agrochemicals and their Remedial Measures

Some Important Terms 184–192

Appendices 193–205

Abbreviations 206–209

Index 209–212

CHAPTER 1

Global Climate Change Impact and Mitigation Strategies

Global climate change means an overall increase in temperature of the universe. The eleventh of the last twelve years (1995–2006) ranks the warmest year in the instrumental records of global surface temperature (since 1850). The 100 year linear warming trend (1906–2006) of 0.7 °C is larger than the corresponding trend of 0.6 °C (1901–2000). The linear warming trend over last 1000 years from 1906 to 2005 (0.13 °C per decades). Observational evidence from all continents and most ocean shows that many natural systems are being affected by regional changes, particularly temperature increase. Climate change means variations in the climate in terms of temperature, relative humidity, sunshine hours, wind velocity and other climatic parameters resulting changes in soil biodiversity, ground water level, soil degradation, erratic and uneven rainfall, frequent droughts and floods. Some common examples of climate change are[2]:

- States like Bihar, Assam, and part of Karnataka are experiencing dry spell, whereas Southern Gujarat, Maharashtara, part of Bihar, Andhra Pradesh, Ladakh and Western Karnataka were hit by the floods.
- In 2007 alone, 17 million people had borne the burnt of floods.
- During the year 2006, the Kashmir valley is witness to most severe summer in three decades.
- Snowfall pattern of the Kashmir valley changes, during January and February no snowfall or less snowfall whereas early snowfall in November and in late March (2008–09 and 2009–10).
- Cherrapunji known for highest rainfall had less rainfall in 2005. Mosinram experiences highest rainfall.
- Mumbai, for consequent 3–4 years, had heavy downpour, almost dipping the city.
- Unusual rainfall (60 cm in 5 days, August 19–23, 2006) in Barmer district of Rajasthan in 2006, was not recorded in the past 200 years.
- Evidences of loss of biodiversity (flora and fauna), genetic materials, soil microorganism at many places.
- Recent tragedy in Kedarnath in Uttrakhand is a lesson for all the Indian states (2013).

- Recent early warning system used at odisha flood is a classical example of use of latest technology, around 1 lakh people have been evacuated. (2013)
- Unexpected snowfall in Kashmir delayed all agricultural activities and fruit crops are subjected to fungal diseases. (2014)

1.1 CAUSES OF CLIMATE CHANGE

Global green house gases (GHGs) emission due to human activities have gone up since pre-industrial times, with an increase of 70 per cent between 1970 to 2004. Carbon dioxide is most important anthropogenic GHGs. Its annual emissions have grown between 1970 and 2004 by about 80 per cent. The largest growth of GHG emission between 1970 to 1994 has come from industry, energy supply, transport, forestry including deforestation, agricultural growth have been decreasing. Global GHGs have increased markedly as a result of human activities since 1750 and now far exceed pre-industrial values determined from ice cores spanning many thousands of years. Several other causes of human made and unforeseen factors functioning together to make situation worsed.

1.2 IMPACT OF CLIMATE CHANGE ON AGRICULTURE AND SYSTEMS

Climatic change is affecting all countries of the world Asian, Southeast Asia or South Asia, European and African in a big way. The poor country would be affected by in a bigger way, however, they are contributing less in climate change. Himalayan ecosystem in which Kashmir valley is situated is not a distant example from the impact of climate change. There is a direct link between the rise of global temperature (1 or 2 °C) and damage to ecosystems. About 130 million hectare land is undergoing different levels of degradation, namely water erosion (32.8 mha), wind erosion (10.8 mha), salinisation (7.0 mha), desertification (68.1 mha), water logging (8.5 mha) and nutrient depletion (3.2 mha)[2]. It has serious impact on the decreasing food productivity due to attack of insect and pest on crop, heavy rainfall, early or late maturity of crop. Small and marginal farmers with small land holding will be more vulnerable to climate change.

Fig. 1.1 a and b (a) View of Kashmir Hillock with snow cover, (b) Rise in temperature induces early maturity in several crops (Photos by Author)

The resilience of many ecosystems is likely to be exceeded this century by an unprecedented combination of climate change and associated disturbances (flooding, drought, wildfire, insects, and ocean acidification) and other global change drivers (e.g. land use change, pollution, fragmentation of natural systems, overexploitation of natural resources). Over the course of this century, net carbon uptake by terrestrial ecosystems is likely to peak before mid century and than weaken or reverse. Approximately 20–30 per cent of plant and animal species assessed so far are likely to be at increased risk of extinction if increase in global average temperature exceeds 1.5 °C to 2.5 °C. At lower altitude, especially in seasonally dry and tropical regions, crop productivity is projected to decrease for even small local temperature increases (1–2 °C), which would increase the risk of hunger. The increase in atmospheric carbon concentration leads to further acidification of atmosphere and earth. Anthropogenic warming could lead to some extended impact depending upon the magnitude of climate change. Uneven and erratic snowfall since last five years had disturbed Himalayan ecosystem.

1.3 CLIMATE CHANGE AND WATER

It is supposed to suppress water resources. On regional scales, mountain snow, glaciers, and small ice caps play a crucial role in freshwater availability. Widespread mass losses from glaciers and reduction in snow cover over recent decades are projecting to accelerate throughout the 21st century, reducing water availability and hydropower potential, and changing seasonality of flows in regions supplied by melt water from major mountain range like Hindukush and Himalayas, where one-sixth of the world population currently lives. Changes in precipitation and temperature leads to changes in runoff and water availability. Runoff is projected to increase by 10–40 per cent by mid century at a higher latitudes and in some wet tropical areas, including populous areas in East and Southeast Asia and decrease by 10–30 per cent over some dry tropics areas due to decrease in rainfall and higher rate of evapotranspiration. Drought affected area are projected to increase in extent, with the potential for adverse impacts on multiple sectors, e.g. agriculture, water supply, energy production and health. Large demand of water in urban areas for domestic purpose is on front and in rural areas for agriculture like irrigation of crops, rarering livestock.

1.4 MITIGATION AND ADAPTATION STRATEGIES FOR COMBATING GLOBAL CLIMATE CHANGE

There should be two fold approaches to mitigate the climate stress—firstly by reducing greenhouse gases (GHGs) emission, the main culprit of climate change and secondly, by adopting necessary farming practices like sustainable forestry systems, diversified cropping systems, carbon sequestration, recourse conserving technologies (RCTs) like zero tillage, minimum tillage or no till system, clean development mechanism (CDM), use of biogas slurry, use of organic manure as a source of plant nutrients and introduction of resistant varieties to droughts, frost, insect pest and logging, etc. Uses of improved farming practices helps in increasing the carbon pool of soils which have been lowered due to overexploitation and soil stress. The CO_2 has been emphasised and overclaimed as compared to NO_2. However N–gases are equally responsible for increasing temperature of atmosphere.

1.4.1 Land Use Practices (LUPs)

Faulty land use practices like shifting cultivation, free-range grazing by cattle, growing crops along with the slope, cultivation of erosion permitting crops, etc. may cause removal of top soil by erosion. Organic matter (OM) has low density than soil solids, hence subjected to easily losses through wind and water erosion. It is clear that the OM loss under 3% slope is around 46 kg/ha in Kerala. Cultivation of soil and consequent aeration stimulate more microbiological activities and promote the oxidation of organic matter, i.e. increase the rate of disappearance of soil organic carbon. Intensive cultivation stimulates decomposition of soil organic matter (SOM). Organic carbon status usually remains low in cultivated soils. It is clear that in all the soil zones, the organic matter content is very high in virgin soil.

1.4.2 Integrated Soil Fertility Management (ISFM)

Over the last few years, the concept of integrated soil fertility management (ISFM) and integrated plant nutrient management (IPNM) has been gaining acceptance. It advocates the careful management of nutrient stocks and flows in a way that leads to profitable and sustained production. The ISFM emphasises management of nutrient flows, but does not ignore other important aspects of the soil complex, such as maintaining organic matter content, soil structure and soil biodiversity. Soil biodiversity reflects the mix and populations of diverse living organisms in the soil — the myriad of invisible microbes to the more familiar macro-fauna, such as earthworms and termites. These organisms interact with one another and with plants and animals forming a web of biological activity. Environmental factors, including temperature, moisture, acidity and several chemical components of the soil affect soil biological activity. Clearly, for a productive sustainable agriculture, the complex interaction among these factors must be understood so that they can be managed as an integrated system.

Soil biota and soil ecosystem health and soil health can be defined as the continued capacity of soil to function as a vital living system, within ecosystem and land-use boundaries, to sustain biological productivity and maintain their water quality as well as plant, animal, and human health[1]. The concept of soil health includes the ecological attributes of the soil, which have implications beyond its quality or capacity to produce a particular crop. These attributes are chiefly those associated with the soil biota; its diversity, its food web structure, its activity and the range of functions it performs. For example, soil biodiversity may not be a soil property that is critical for the production of a given crop, but it is a property that may be vital for the continued capacity of the soil to produce that crop. The soil biomass is considered as an index of soil activity for good crop growth and productivity.

1.4.3 Biological Management of Soil Fertility (BMSF)

It is a central paradigm for the biological management of soil fertility to utilise farmer's management practices to influence soil biological populations and processes in such a way as to achieve desirable effects on soil productivity[1]. Biological populations and processes influence soil fertility and structure in a variety of ways, each of which can have an ameliorating effect on the main soil-based constraints to productivity: symbionts, such as rhizobia and mycorrhiza increase the efficiency of nutrient acquisition by plants; a wide range of fungi, bacteria, and animals participate in the process of decomposition, mineralization, and nutrient immobilisation and therefore influence the efficiency of nutrient cycles; soil organisms mediate both the synthesis and decomposition of soil organic matter and therefore influence cation exchange capacity, the soil N, S, and P reserve, soil acidity and toxicity; and soil water holding capacity; the burrowing and particle transport activities of soil fauna, and the aggregation of soil particles by fungi and bacteria, influence soil structure and soil water regime.

1.4.4 Role of Biodiversity

The role of soil biota/biodiversity in sustaining the productivity of agricultural systems. A fundamental shift is taking place worldwide in agricultural research and food production in climate change scenario[1]. In the past, the principal driving force was to increase the yield potential of food crops and to maximise productivity. Today, the drive for productivity is increasingly combined with a desire and even a demand for sustainability[1]. Sustainable agriculture involves the successful management of agricultural resources to satisfy human needs while maintaining or enhancing environmental quality and conserving natural for future generations. Improvement in agricultural sustainability will require the optimal use and management of soil physical properties. Both rely on soil biological process and soil biodiversity. This implies management practices that enhance soil biological activity and thereby build up long-term soil productivity and health. Such practices are of major importance in marginal lands to avoid degradation, in degraded lands in need of restoration and in regions where high external input agriculture is not feasible. Recently a national conference on biodiversity organised by KU to discuss several key problems of biodiversity.

1.4.5 Resources Conserving Technologies (RCTs)

RCTs are very important for increasing the carbon pool in soils. Zero tillage system offers minimum soil disturbances during sowing of crops. Zero tillage, raise bed planting, cover crops, crop residue management and mulching proves unique opportunities for restoration of soil organic carbon in agricultural lands. Combating climate change in time, it is imperative to use renewable energy sources at domestic level. The uses of refrigerators shall be restricted to reduce the emission of gases. Carbon sequestration by composting, rising of green legumes and uses of manure is an important activity in agricultural production systems. Effort must be made at all levels of societies, institutions, NGOs and other youth forms. People must sensitise to use less combustive vehicles, preference shall be given to CNG (compressed natural gases) vehicles. The use of renewable energy sources like solar torch, solar batteries, solar water heating system must be recognised for energy saving and reducing GHG emission. Laser land leveling have been proved beneficial for water saving and equal distribution of soil nutrients of water for plants and fruit crops.

1.5 KEY REFERENCES AND RESOURCES FOR FURTHER READING

1. FAO programme on agriculture biodiversity, soil biodiversity portal at www.fao.org/ag/ag1/ag11/soilbiod/fao.htm
2. Agarwal KP, Climate Change and its Impact on Agriculture and Food Security, *Leisa India*, 2008; 10:4:6–7.

SOME GREAT THOUGHTS

- Soil is soul of infinite lives. —LL Somani
- One should be man of values rather than man of success. —Albert Einstein
- Light for some day is good but continuous light is exhaustive. —Alexander von Humboldt
- FAIL means first attempt, initial lesson to work hard. —APJ Abdul Kalam
- Use of organic inputs save our ecosystems, soil and plant biodiversity at greater extent in Himalayas and Hindu Kush Himalaya (HKH). —Tej Pratap
- Quality seed is an important input for high productivity. —SA Wani

CHAPTER 2

Soil Health (Quality) *vis-à-vis* Soil Organic Carbon and Food Security for Social Development

Soil quality (SQ) is capacity of specific kind of a soil to function, within natural and managed ecosystem boundaries, to sustain plant and animal productivity, maintain or enhance water and air quality and support human health and habitation[4]. In the mid sixteen due to introduction of dwarf varieties, which need high input in term of fertilizers, water, pesticides increased agricultural production, but discriminate and imbalance use of fertilizer deteriorate inherent capacity of soil to supply plant nutrient. Soil health (SH) refers to the fitness of soil for any specific purpose determined by the factors chosen for soil classification, soil suitability and land capability. It examines spatial and temporal variations induced by land use policy or management. Soil organic carbon (SOC) is the most reliable, versatile and easily assessable indicator, encompassing interactive effect of several factors. Plateauing or decreasing trends of crop yields at current level of management indicate declining SH. Erosion, drought and desertification, irrigation induced salinity and sodicity, paradigm shift in land use, nutrient depletion and intensive cultivation are the cause of SH deterioration. Erratic rainfall and exploitation of land, water and vegetation resources by ever increasing human and livestock population further accentuate the problem of SH. Increasing salinity, residual carbonate, alkalinity and contamination of surface and ground water through heavy metals, nitrates, fluoride and arsenic are the reflection of deteriorating soil health (SH).

The soil health is an indicator of good soil physical, biological and chemical properties for higher production. The challenges of growing population, industrialization and urbanization on qualities of diminishing resources are quite daunting especially in Asia, Africa and other developing nations. Soil, water and biodiversity are integral part of sustainable production system in the era of resource degradation and heavy input depended agriculture. In the mid sixteen our natural resources were not exploited, now have been reached up to the high level of exploitation, need immediate attentions. Enhancing productivity by input intensive agriculture consolidated food and nutritional securities of developing nations with limited per capita availability of land, water, and bio resources[6]. As per recommendations made in proceedings of International Conference on Soil, Water and Environmental Quality-Issues and Strategies (ICSWEQ) held from January 28 to February 1, 2005 at New Delhi, the shrinking capacity of soils to absorb any more abuse must be impressed in the public mind through appropriate changes in educational curriculum, mind set, awareness, mass media and it is the time for individual countries to act on the "World Soil Charter of FAO" and press for "UN Soils Convention" to accord the same high priority to soil preservation as is being currently given to climate change, biodiversity, etc. SOM is the mainstay of soil quality. While balanced fertilization may meet crop productivity and maintain SOM. It is an urgent imperative to improve the sequestration of carbon in all the soils by all available resources including recycling of crop residues, green manuring, composting, zero tillage, RCTs and other soil agro techniques. (Subhash *et al.*, 2013)

2.1 CHALLENGES OF SOIL QUALITY

The mother earth day is celebrated every year on 22nd April worldwide for creating awareness about the challenges of land degradation, soil pollution, and climate change impact and for earth preservation. Against an annual depletion of 28 million tonnes (mt) of nutrients, against addition of 20 mt, leaving a net gap of 8 mt per annum, a deficiency which is accumulating year after year, depleting SQ. A fact finding committee constituted by the Government of India in 1997 with the objective of analyzing in-depth the trend of productivity of important crops in Haryana and Punjab reported that there is decline in the organic-carbon content of the soils due to continuous cultivation of cereal based cropping system for instance rice-wheat, rice-rice and rice-maize, etc. The continuous nutrients depletion from the agriculture field is the severe threat to the SH. Sub-optimal nutrient application together with poor quality water results in sparse plant cover and low vegetative inputs into the soils. In arid and rain-fed areas removal by crop was far more than the added through fertilizers[6]. Alfisols, ultisols and oxisols with low cation exchange capacity (CEC) were the heaviest looser. Soil organic carbon loss ranged from 0.22 to 6.0% over the initial in different cropping sequences of India due to inadequate fertilization, whereas depletion of SOC was far less 0.22 to 2.92% under balanced fertilization[1]. Inadequate fertilization with high RSC (residual sodium carbonate) depleted phosphorus and potassium by 7.7 and 13.4%, respectively in arid soils of India from 1975 to 2002[9]. The abrupt change in land use by introducing high water requiring crops further heighten the problem of nutrient depletion. Traditional farming system was extensive with low yields, which was sustainable in harmony with the carrying capacity set by the nature. Low productivity system has lost relevance in view of increasing demand of food, fiber and wood. Since 1951–52 there has been an increase of 36, 22 and 54 million hectare irrigated, net sown and double-cropped area, respectively on the cost of fallow, pastures and grazing lands and tree grooves. High intensity farming system supports high productivity, they appear non-sustainable in absence of holistic land management which satisfies needs of stakeholders in an economically favorable way and simultaneously contains curative action for preserving the SH and prevents further soil degradation.

Due to concerns with soil degradation and the need for sustainable soil management in agro ecosystem, there has been much scientific attention to characterize soil quality. The Soil Science Society of America (SSSA, 1995) defines SQ as *"it is the capacity of a specific kind of soil to function within natural or managed ecosystem boundaries, to sustain plant and animal productivity, maintain or enhance soil and water quality, and support human health and habitation*. In simplest terms, SQ or SH can be defined as *"the fitness of soil for use"*. In agricultural systems, high quality soil provides for the sustained and productive growth of crops with minimal impacts on the environment. There are two components of SQ viz. inherent and dynamic. Inherent SQ refers to the characteristics that define a soil's inherent capacity for plant production. These are usually static, changing little over short time frames (years to decades). Soil texture and soil mineralogy are commonly included as properties of inherent SQ. Other soil properties, such as total soil carbon, CEC and exchangeable sodium percentage (ESP) may also be defined as inherent properties where broad soil type or regional comparisons are of interest at one point of time, even though they may be altered by management over longer time frame. Inherent soil properties are the basis of many land use capability or suitability assessments that are key components of land use planning and policy development in many regions. Most were undertaken with the primary aim of evaluating potential soil productivity. Thus, the inherent quality of a soil should be viewed in light of its intended agricultural use. Properties of dynamic SQ are those that change in response to human use and management normally over relatively short time frame (years to decades). Agricultural soils of high dynamic SQ maintain high nutrient availability, permit adequate infiltration of water and air, have relatively stable structure and maintain a functionally diverse community of soil organisms that support a relatively high level of plant productivity. These processes are reflected in specific physical, chemical and biological properties of soils. The terms "dynamic soil quality" and "soil health" are often used interchangeably. Two soils may be equally "healthy" but achieve different levels of plant productivity because of differences in their inherent quality.

2.2 CHALLENGES OF FOOD AND SOCIAL SECURITY

Food an individual eats fundamentally affects his health, strength, stamina, nervous condition, moral and mental functioning. It is of paramount importance in the normal growth, development and health of humans. The access to food by all is still unachieved but cherished goal. A widely accepted food security comprises of three parts. Every individual has a physical, economic, social and environmental access to balanced diet that includes necessary macro and micronutrients, safe drinking water, sanitation, environmental hygiene, primary health care and education so as to lead a healthy life. Food is produced from efficient and environmental friendly technologies that conserve and enhance the natural resource base of crops, animal husbandry, forestry, inland and marine fisheries, etc. The ultimate composition of a food is expressed in terms of nineteen chemical elements.

Every human being has fundamental rights of balanced and nutritious diets in required quantity every day for sustaining their lives. Food security is a distance goal but achievable. Projections made by Food and Agriculture Organization (FAO) on trends in agriculture, food and forestry (*World Agriculture: Towards 2015/2030*) indicate that low income countries with high dependence on agriculture will encounter challenges of, food security, sustainability and rural poverty. In the past, the increasing needs of expanding population for food, fuel and fiber were met from cultivating progressively larger areas of land and by intensifying the use of existing cultivated land. Under the circumstances when no more additional good quality land is available and the crop yields are stagnated, the food requirement of added population in future has to come from the reclamation and management of degraded lands which include salt affected ones also. India has world's 2.4% of land and 4% of fresh water resources of which nearly 6.73 million hectares lands are salt affected and a sizeable area is underlain by poor quality water. With these limited resources we have to support 16% of the global population. Owing to higher allocation of good quality water to other remunerative sectors the availability of fresh water to agriculture which is the largest user, is diminishing rapidly and has to largely depend upon the low quality water resources. Therefore, another important component to achieve higher productivity could be the optimum utilization of surface waters as well as low quality groundwater and waste waters, which are still usable. The injudicious use of water is often associated with the development of water logging, salinity, sodicity and many other environmental problems. Adequate knowledge in diagnosis and management technologies for saline and alkali lands/waters and wastewater generated from municipalities and industries is essential to obtain maximum crop production from these resources for achieving goal of 300 million tonnes food grains by 2020 from 145 mha Indian arable lands. The social security depends upon equal distribution of services.

2.3 ENHANCING SOIL ORGANIC CARBON (SOC) STOCK — AN ULTIMATE SOLUTION

The SOM consists of living organisms (bacteria, fungi, earthworms, nematodes, insects, and plant roots), active organic matter (fresh/partially decomposed, labile) and humus (well decomposed and relatively stable). The source of organic matter (OM) in soil is plant and animal residues and the products synthesized by them and microorganisms. The OM and humus serve as reservoirs of living organisms and these living organisms participate in the mobilization of plant nutrients and facilitate to build soil structure besides providing other benefits. Nutritional value of organic matter lies in its dynamic nature. Organic carbon is the energy source for soil organisms, and it is the activity of these organisms and the processes they are involved in rather than the absolute-organic matter level, which is most important. Increasing SOC in tropical climate is not easy. However, continuous application of lingocellulotic crop residue helps in building soil organic matter temporarily. There are evidences in literature, as shown through long term experiments, that application of farm yard manure (FYM) + Mineral nutrient/NPKS (nitrogen, phosphorus, potash and sulphur) and micronutrient increases the crop yield as well as build the organic carbon in soils. Integrated use of manures and fertilizers helps in building SOC residues which have a major role in maintaining soil organic matter content. Soil microorganisms grow rapidly during early phases of decomposition when plant materials are subjected to the transformation and decomposition processes by the heterotrophic microflora immediately after incorporation into the soil, and as a result, the population of bacteria, fungi and actinomycetes increased with application of plant residues and FYM. It was also observed that exhaustion of the available carbon led to decrease in microbial biomass. Many crop management practices, such as manuring including green manuring, crop residue application, mulching have shown improvement in soil organic matter and microbial population.

Table 2.1 Post harvest buildup of SOC by uses of different organic residues under different cropping sequences[8]

Land use systems	**Material added**	**Organic carbon (%)**
Maize-wheat (25 yrs)	Control	0.51
	FYM	2.49
Cotton-sorghum (45 yrs)	Control	0.56
	FYM	1.14
Ragi-cowpea-maize (3 yrs)	Control	0.30
	FYM	0.64
Rice-rice (10 yrs)	Control	0.43
	50% from inorganic + 50% through green manuring (*Sesbania aculeate*)	0.90
Rice-wheat (3 yrs)	Control	0.44
	FYM	0.54
Rice-wheat (7 yrs)	Fellow	0.23
	Green manuring (*Sesbania aculeate*)	0.37

Soil carbon sequestration refers to the storage of carbon into a stable solid form. It occurs through direct and indirect fixation of atmospheric CO_2. Direct soil carbon sequestration occurs by inorganic chemical reaction that converts CO_2 into soil inorganic carbon compounds, such as Ca and Mg carbonates. Indirect plant carbon sequestration occurs during plants photosynthesis in which atmospheric CO_2 is converted into plant biomass; subsequently some of the plant biomass is indirectly sequestered as soil organic carbon during decomposition process. The amount of carbon sequestered at a site reflects the long term balance between carbon uptake and release mechanism. Many best management practices have been proven to help in sequestering soil carbon like restoration of degraded soils and ecosystems. The adoption of recommended agricultural practices on prime land and retiring marginal agricultural lands to restorative land uses or converting to natural ecosystems. With rapidly increasing population restoration of degraded soils and ecosystems is an important strategy. This strategy of restoration of degraded soils and ecosystems can enhance biomass production, improves soil quality and increases the SOC pool. Many soils of the tropics especially those in densely populated regions of Asia have lost a large proportion of their original SOC pool because of practices of mining soil fertility. There is a large potential of restoration of degraded soils in Southeast Asia which ranges from 18.3 to 35.0 teragram carbons per year (TgC/yr). These estimates are attainable potentials provided that regional governments adopt appropriate polices and implement plans to restore degraded soils through forestation, establishing planted fallows and improving grazing lands. It is a major challenge that must be addressed in a coordinated and planned manner.

The SOM is extremely important for productivity, and particularly so for the poorer soils of arid and semi-arid areas. Its direct contributions to nitrogen and sulphur nutrition of crops, and its role in stabilizing soil aggregates and supporting the soil biota responsible for creating pores through which air and water move cannot be ignored. In addition, soil organic matter plays a major role in the retention of cationic nutrients by dominant soils of these areas which have clays composed of kaolinite, and low activity iron and aluminium oxides clays with only a weak ability to hold nutrient cations. Furthermore, under acid conditions, some of the organic compounds present in soil form complexes with aluminium which would otherwise be toxic to plants. In addition to physical and chemical effects, organic matter provides substrate for supporting biological life in the soil.

Table 2.2 The changes in soil properties (0–30 cm) under different tree-crop combinations in 5 years[8]

Land use system	**Organic carbon (%)**	**Available N (kg ha^{-1})**
Crop based system	+0.07	+10
Eucalyptus based	+0.12	+21
Acacia based	+0.20	+31
Populus based	+0.17	+25

Under natural vegetation the amount of organic matter in the soil tends to be established at a relatively high level, but under cultivation, addition is usually much less than from the natural vegetation, and consequently the OM level tends to fall. If good crops are grown and all residues returned to the soil, the level established after cropping may be different than under grassland. A general principle of sustainable SH management systems is that return as much organic material as possible to arable upland soils, but it should be free from toxic contaminants, and the costs and problems of collecting and spreading should be socially and economically acceptable.

The ISCA 101st online science congress recently organised at Jammu, concluded "that people's mindsets are changing very fast toward scientific innovations. Science makes people more vibrant for soil quality and soil health restoration. All sciences enable humans to fair future challenges."

2.4 KEY REFERENCES AND RESOURCES FOR FURTHER READING

1. Anonymous, Annual Reports of NATP for Rainfed Agro Ecosystem. Central Institute of Dry Land Agriculture, Hyderabad, India 2002–03; p124.
2. Yadav G, Lal Khajanchi. Management of Land Degradation in Arid and Semi Arid Areas. In Diagnosis and Management of Poor Quality Water and Salt Affected Soils. Eds. Lal K, Meena RL, Gupta SK, Saxena CK, Singh Gajender, Singh Gurbachan. Central Soil Salinity Research Institute, Karnal, India 2008; p311.
3. Goswami NN. Soil and its Quality *vis-à-vis* Sustainability and Society: Some Random Thoughts. In: Proceedings of International Conference on Soil, Water, and Environment Quality-Issues and Strategies. Published by Indian Society of Soil Science 2005; p43–58.
4. Karlen DL, Mausbach MJ, Doren JW, Cline RG, Harris RF, Schuman GE. Soil quality: A Concept, Definition and Framework for Evaluation. *Soil Science Society of America Journal* 1997; 6:4–8.
5. Katyal JC. Soil Fertility Management—A Key to Prevent Desertification. *Journal of the Indian Society of Soil Science* 2003; 51:378–487.
6. Samra JS. Participatory Watershed Management for Improved Soil and Water Quality. In: Proceedings of International Conference on Soil, Water, and Environment Quality-Issues and Strategies. Published by Indian Society of Soil Science 2005; p22–29.
7. Chand S. Integrated Nutrient Management for Sustaining Crop productivity and Soil Health. International Book Distributing Company, Lucknow, India 2008; p112.
8. Swarup A, Manna MC, and Singh GB. Impact of Land Use and Management Practices on Organic Carbon Dynamics in Soils of India. In: Global Climate Change and Tropical Ecosystems. Eds. Lal R, Kimble JM, Eswaran H, Stweart BA. Lewis Publishers, Boca Raton, Fl 1999; p261–281.
9. Singh SK, Kumar Mahesh, Sharma BK, Tarafdar JC. Organic Carbon, Phosphorus, and Potassium Depletion Under Pearl Millet based Production System in Arid Rajasthan. Arid Land Research Management 2007; 21:119–131.
10. Chand S, S Lal and Singh P. Sustainable Agriculture, Food Security and Climate Change. Daya Publishing House, New Delhi, 2012; p218.

CHAPTER 3

Carbon Sequestration — A Solution for Mitigating Global Climate Change/Global Warming

Recently released on report of WG of IPCC on 31st March, 2014 clearly indicated several hazards and risk for human life, due to increase in temp. Global warming is an increase in average measured temperature of the earth's surface and oceans since the mid-twentieth century. The average global air temperature near earth's surface increased 0.5°–1.0 °F during the last 100 years. The Inter Governmental Panel on Climate Change (IPCC) concluded that average global surface temperature is likely to rise a further by 1.4 to 5.8 °C or even warmer during the 21st century. This increase in temperature is well correlated with the anthropogenic causes especially with the increase in concentration of different GHGs (CO_2, CH_4, N_2O)[10]. Concerns regarding global warming and increase GHG concentrations have led to a question about the role of soils as a carbon source or sink[7]. Soils constitute the largest surface C pool, approximately 1500 gigatonne C which is 3 times the quantity stored in terrestrial biomass and twice that in atmosphere[14]. Therefore, any modification of land use and management practise can change soil carbon stocks[16].

Conservation technologies coupled with INM are important to conserve soil organic carbon in an cultivated land. Soil carbon sequestration refers to the storage of CO_2 into stable form so that it is not immediately remitted[12]. In India the present soil organic carbon (SOC) pool of 329 mha soils is 24.3 pentagram (Pg) with a potential SOC capacity of 34.9 Pg[4]. While global potential of soil carbon sequestration is 1–2 Pg C/yr[10]. Thus, carbon sequestration is a bridge to the future and a natural, cost effective and eco-friendly option of cleaning dirty atmosphere. This review article highlights the opportunities of soil carbon sequestration and its importance for mitigating global climatic change in agriculture.

Global climate change is the most serious environmental problem of 21st century[7]. This change in climate has already begun with a global temperature increase of 0.5°–1.0 °F (0.6 °C) since 20th century. In the 20th century 12 warmest years were recorded and all occurred in the last 15 years of the century of those 1998 was the warmest year on record. The snow cover in the northern hemisphere and fronting ice in the Arctic Ocean have decreased, and globally sea level has risen by +0.1–+.2 m over the past century. Also change in precipitation up to +0.5–1.0%/

decade was recorded. As per IPCC the global mean temperature may increase between 1.4 and 5.8 °C by 2100.

The increase in temperature is well correlated with anthropogenic activities and especially with the increase in the concentration of green house gases (GHGs). Out of three major GHGs, CO_2 is the major contributor to the global warming. There is an increase in the atmosphere CO_2 concentration by 31% since 1750 from fossil fuel combustion and land use changes. This global warming caused as a result of increased GHG emissions especially that of CO_2 will result in following rapid effect[9]:

1. Large scale disruption of forestry, agriculture and fisheries
2. Extinction of many plant and animal species in land and oceans
3. Changing rainfall and snowfall pattern
4. Lost of huge tracts of coastal lands under rising seas as the oceans expand due to melting of polar ice
5. Less assess to less reliable water supplies in many parts of the world
6. Serious adverse effects on human health

All these above mentioned effects necessitate the identification of strategies for mitigating the thrust of the attendant global warming. There are two principal strategies of mitigating the green house effect.

1. Reduction in emission of greenhouse gases
2. Sequestration of atmospheric CO_2 into biomass and soil

Limiting GHG emissions is not enough to avoid dangerous climate change as too much carbon is already in the atmosphere. The IPCC stated in 2007 that complete elimination of CO_2 emissions is estimated to lead to a slow decrease in atmospheric CO_2 of about 40 ppm (part per million) over 21st century. In other words, the strategy of reducing CO_2 emission by itself has little effect on the atmospheric concentrations. Thus there is a biological side also (i.e. carbon capture or sequestration system) which could store, convert or recycle GHGs preventing them from building up the atmosphere. (Subhash Chand *et al.*, 2012)

Also residence time of C in soil is 35 years compared to 5 years in atmosphere and 10 years in vegetation. Hence, it is apparent that soils play a significant role in the control of the C stocks and fluxes. Thus with regard to the potential of the soils to mitigate the GHG affect and more generally with regard to land use change and forestry the term soil carbon sequestration comes into play.

3.1 SOIL CARBON SEQUESTRATION (SCS)

Soil carbon sequestration refers to the storage of C into a stable solid form. It occurs through direct and indirect fixation of atmospheric CO_2. Direct soil C sequestration occurs by inorganic chemical reaction that converts CO_2 into soil inorganic carbon compounds, such as Ca and Mg carbonates. Indirect plant C sequestration occurs during plants photosynthesis in which atmospheric CO_2 is converted into plant biomass; subsequently some of the plant biomass is indirectly sequestered as soil organic carbon during decomposition process. The amount of carbon sequestered at a site reflects the long term balance between carbon uptake and release mechanism. Many best management practices have been proven to help in sequestering soil carbon are given below:

1. Restoration of degraded soils and ecosystems
2. Adoption of recommended agricultural practices on prime land
3. Retiring marginal agricultural lands to restorative land uses or converting to natural ecosystems

In Haryana, India, it is reported a large increase in SOC content by reclamation of sodic soils through growing *Prosopis juliflora* (Table 3.1) The biomass production increased to 318 megagram per hectare (Mg/ha) above and 44 Mg/ha below ground with a mean value of 12.1 Mg/ha/yr.[3]

Table 3.1 Soil carbon sequestration through restoration of degraded soils

Degradation process	Area (mha)	Rate of soil organic carbon sequestration (kg C/ha/yr)	Total potential (Teragram C/yr)
Water erosion	81.2	100–150	8.1–12.2
Wind erosion	10.8	20.50	0.2–0.5
Desertification control			
1. Irrigated land	16.1	100–150	1.6–2.4
2. Rainfed	67.0	20–50	1.3–3.4
Salinization	33.0	200–500	6.6–16.5
Fertility depletion	10.9	100–150	1.1–1.6
Total			18.3–35.0

Source: (9)

Data in Table 3.2 shows drastic increase in the SOC pool through restoration with an annual rate of increase of 0.3 MgC/ha/yr in first 5 years, 3.6 MgC/ha/yr from 5 to 7 years, 1.6 MgC/ha/yr from 7 to 30 years and 1.4 MgC/ha/yr as an average rate for the entire 30 year period. Similar restorative measures could be adopted on soils degraded by the other process like erosion, compaction, etc.

Table 3.2 Soil organic carbon sequestration by growing prosopis in a sodic soil in Karnal

Depth (cm)	SOC pool in years after planting (Megagram C/ha)			
	0	5 yrs	7 yrs	30 yrs
15	3.5	5.0	14.3	21.5
30	3.5	3.5	7.2	10.1
60	2.7	2.7	7.4	10.8

Contd.

Table 3.2 Soil organic carbon sequestration by growing prosopis in a sodic soil in Karnal *(Contd.)*

Depth (cm)	SOC pool in years after planting (Megagram C/ha)			
	0	5 yrs	7 yrs	30 yrs
90	1.6	1.6	3.7	8.3
120	0.5	0.5	1.6	3.6
Total	11.8	13.3	34.2	54.3

Source: (3)

Phosphate mines in Northern and Central Florida provide a valuable resource for the national and international production of agricultural fertilizers. However, separating phosphate rich ore from the underlying sand and clay matrix creates large containment ponds or clay settling areas (CSA). The physical and chemical characteristics of CSA lands makes restoration a critical priority for post mining activities. Therefore, to demonstrate the potential use of these degraded areas for bioenergy crop production and carbon sequestration, 50 ha demonstration planting consisting of *Eucalyptus grandis* and cogongrass was established near Lakeland, Florida[15]. Soil carbon measurement for soils sampled from beneath cogongrass and *E. grandis* was carried out and both the concentration and stock of carbon was higher in the upper 10–20 cm of soil and lower in 40–50 cm (Table 3.3).

Table 3.3 SOC measured at two depths for cogongrass and E. grandis on clay settling areas in Florida

Soil property	Soil depth (cm)	Cogongrass	*E. grandis*	% change
Carbon%	10–20	1.68	5.29	+215
	40–50	0.76	3.06	+303
Carbon stock	0–30	7.1	22.2	+214
(kg/cm)	30–60	3.2	12.8	+304

Source: (17)

3.2 ADOPTION OF RECOMMENDED AGRICULTURAL PRACTICES (RAPs)

Adoption of recommended agricultural practises (RAPs) can enhance the production of above and below ground biomass and increase the SOC content. In comparison with other land uses, e.g. permanent pastures and forests, crop lands have lost most of their original SOC pool because of ploughing and susceptibility to erosion. Thus, there is a tremendous scope for improving SOC content through adoption of RAPs on crop lands. In addition to growing high yielding varieties within appropriate cropping system other components of RAPs are given below:

1. Using conservation tillage
2. Soil fertility management
3. Mulching/cover crops
4. Enhancing rotational complexity

If the above mentioned RMPs or components are followed then the South Asian countries have a potential of SOC sequestration equal to 11 to 22 teragram carbon per year (TgC/yr) of which 8–16 TgC/yr is in cropland of India (Table 3.4)

Table 3.4 Carbon sequestration potential through adoption of RAPs in South Asia

Country	Area (mha)	Rate of SOC sequestration (kg C/yr/ha)	Total potential (Teragram C)
Afghanistan	7.9	50–100	0.1–0.2
Bangladesh	8.1	200–300	1.2–1.8
India	161.8	100–200	8.1–16.2
Iran	14.3	50–100	0.4–0.8
Nepal	3.1	300–500	0.7–1.2
Pakistan	21.5	50–100	0.5–1.0
Sri lanka	0.9	300–500	0.2–0.4
Total	–	–	11.2–21.6

Source: (1)

The INM practices are most reliable tools for carbon sequestration and crop and soil management in modern agriculture. Chand[5] compiled 200 INM practices for sustainable crop productivity and soil health. This drastic decrease in the C content is explained by the so-called de-protection of soil organic matter physically protected inside soil aggregates due to the heavy tillage operations[1].

Changes in agricultural management can increase or decrease SOC. With the aim a study was conducted by West and post to quantify potential soil C sequestration rates for different crops in response to decreasing tillage intensity and by enhancing rotation complexity. The results (Figs 3.1 and 3.2) indicate that on an average the change from conventional tillage (CT) to No-till (NT) can sequester 57 ± 14 grams $cm^{-2}yr^{-1}$ while enhancing rotational complexity can sequester on an average 20 ± 12 gm $cm^{-2}yr^{-1}$. Also carbon sequestration rates with a change from CT to NT can be expected to peak in 5 to 10 years with soil reaching a new equilibrium in 15 to 20 years. While by enhancing the rotational complexity a new SOC equilibrium will reach in approximately 40 to 60 years. The soil organic carbon is the central atom for all activities and for improving soil productivity by providing energy to soil microbes for respiration and feeding. (Subhash 2010)

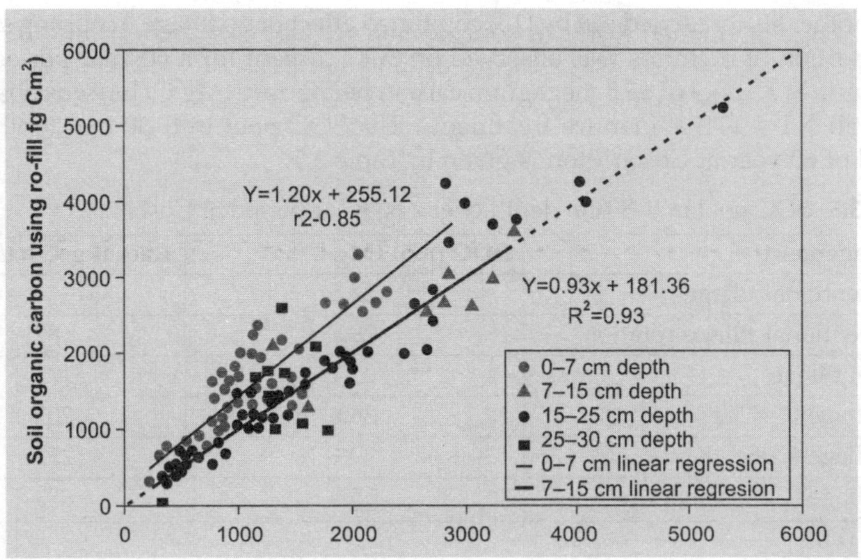

Fig. 3.1 Carbon sequestration rates with a change from CT to NT

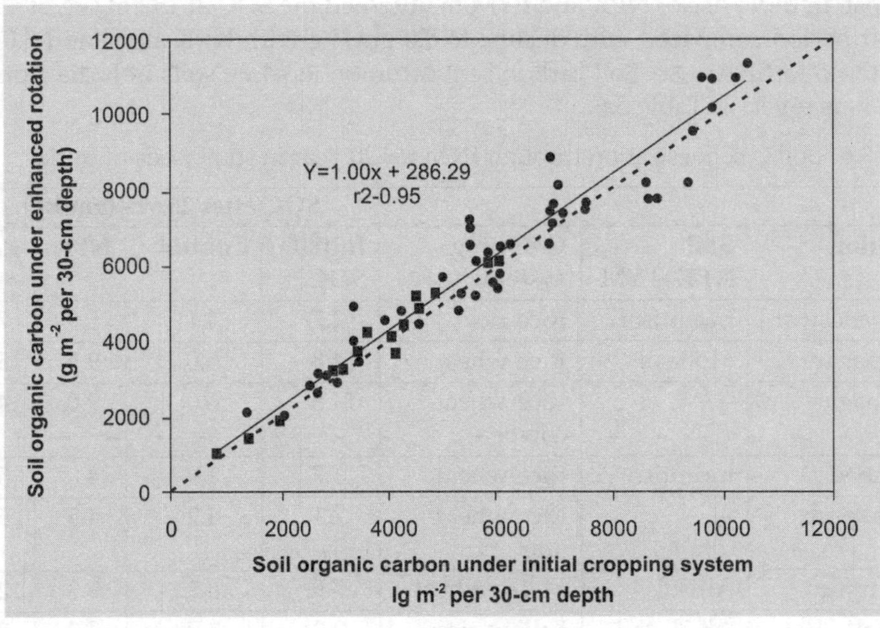

Fig. 3.2 Carbon sequestration rates with a change of crop rotation

In another study carried out by (13) combined effect of no tillage, cropping system and addition of manures was observed on SOC content for a 60 year period. The maximum SOC pool of 65.5 megagram carbon per hectare (Mg C/ha) was observed in no till NT + C-S + manure treatment. The SOC pool in 0–30 cm depth over period of 60 year at Coschocton is given in Table 3.5.

Table 3.5 SOC pool in 0–30 cm depth over a 60–year period at Coshocton

Management	SOC pool (Mg C/ha)	Rate (kg C/ha/yr)
Conventional tillage	24.5	–
Conventional tillage-rotation	29.7	87
Chisel tillage	32.1	127
No tillage (C–C)	36.8	205
No tillage (C–S)	39.6	252
No tillage (C–S) + manure	65.5	683

Source: (13)

The SOC pool can also be increased by the addition of crop residue, maintenance of soil fertility through the integrated application of fertilizers, cattle manure and compost. In India use of NPK and FYM maintained the SOC at 15 gm/kg of soil for 25 year period compared with decline to 8.0 gm/kg with NPK alone and 5.0 gm/kg with no fertiliser use. Soil carbon sequestration in some soils in India from last 20 years is given in Table 3.6.

Table 3.6 Soil C sequestration through INM for 20 year in some soils of India

			SOC after 20 yrs (gm/kg)			
Location	**Soil NPK+FYM**	**Cropping system**	**Initial SOC**	**Control**	**NPK**	**g/kg**
Bhubaneswar	Inceptisol	Rice-rice	2.7	4.1	5.9	7.6
Pantnagar	Mollisol	Rice-wheat	14.8	5.0	9.5	15.1
Pantnagar	–	Rice-wheat-cowpea	14.8	6.0	9.0	14.4
Faizabad	Inceptisol	Rice-wheat	3.7	1.9	4.0	5.0
Barrakpore	–	Rice-wheat-jute	7.1	4.2	4.5	5.2
Palampur	Alfisol	Maize-wheat	7.9	6.2	8.3	12.0
Karnal	Alkali soil	Fallow-rice-wheat	2.3	3.0	3.2	3.5
Nagpur	Vertisol	Cotton-cotton	4.1	–	–	5.5
Trivandrum	Ultisol	Cassava	7.0	2.6	6.0	9.8

Source: (18)

The locations data in Table 3.6 shows increase in SOC content by application of NPK and manure in 6 out of 9 soils and increase in 3 locations out of 9 locations soils by use of NPK. There was a drastic decrease in SOC content in all soils which received neither fertiliser nor manure.

In the Lingyou and Qingan region of China a survey was conducted to see the effect of land use on carbon stocks in soils. The carbon content ranged from more than 40 megagram carbon per hectare (Mg C/ha) in forest soils, 60.0 Mg C/ha in grassland and 12.1 Mg C/ha in cropland of Lingyou region while as in Qingan region OC ranged from 31.8 Mg C/ha in forest soils, 16.3 Mg C/ha in grasslands and 14.4 Mg C/ha in croplands. The low values of OC in croplands are explained by the significant loss of carbon that had occurred during soil cultivation[21]. The soil carbon sequestration in China is given in Table 3.7.

Table 3.7 The SOC in various land use systems for Lingyou and Qingan sites in China

Location	Total SOC pool (Mg C/ha)			
	Depth (m)	**Forest**	**Grassland**	**Cropland**
1. Lingyou	0–0.1	33.2	40.3	9.4
	0.2–0.3	6.9	19.7	2.7
Total		40.1	60.0	12.1
2. Qingan	0–0.1	23.3	13.1	12.1
	0.2–0.3	8.5	3.2	2.3
Total		31.8	16.3	14.4

Source: (21)

In another experiment carried out a slow increase in SOC was observed from 35,000 to 75,000 lb C/acre (Fig. 3.3) in a period of 100 years after planting a depleted marginal land with pine trees[2].

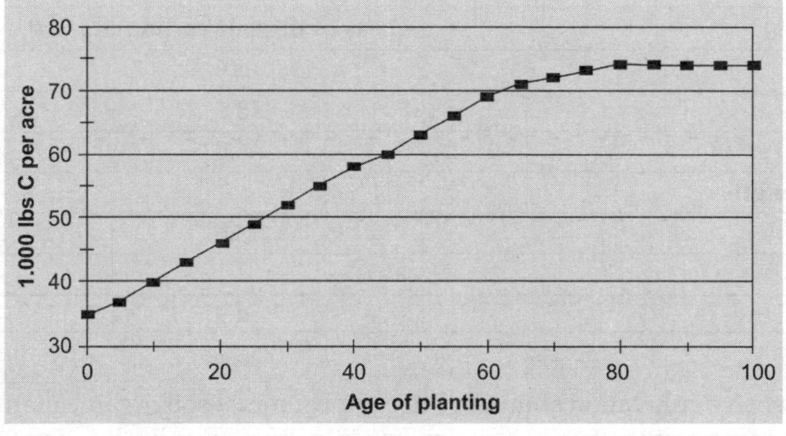

Fig. 3.3 Increase in SOC with increasing age of pine trees

3.3 LAND USE PRACTICES (LUPs)

The cultivation of soil and consequent aeration stimulate more microbiological activities and promote the oxidation of organic matter, i.e. increase the rate of disappearance of soil organic carbon[4]. The current and potential SOC capacity of Indian soils is given in Table 3.8. Out of total area of 328.5 m/ha, current SOC pool is 24.3 pentagram carbon (Pg C) but at the same time Indian soils have got potential of sequestering 34.9 Pg C while global potential of soil carbon sequestration is 1–2 pg C/yr which is equal to 24% of the total emissions by fossil fuel combustion or talking in terms of fossil fuel, this global potential of 1–2 Pg C/yr is equal to 1–2 billion barrels of diesel per year.

Table 3.8 Current and potential SOC capacity of Indian soils

		SOC (pentagram C) pool	
Soil type	**Area (mha)**	**Current**	**Potential**
Red soil	84.6	6.8	10.0
Black soils	98.8	9.9	13.4
Alluvial soils	103.8	5.0	7.7
Mountainous soils	41.3	2.6	3.8
Total	328.5	24.3	34.9

Source : (4)

The faulty land use practices like shifting cultivation, free-range grazing by cattle, growing crops alongwith the slope, cultivation of erosion permitting crops, etc. may cause removal of top soil by erosion[20]. Organic matter has low density than soil solids hence subjected to easily losses through wind and water erosion. It is clear that the OM loss under 3% slopes is around 46 kg/ha in Kerala (Table 3.9).

Table 3.9 Organic carbon losses under different slopes and slope length

Slope (%)	**Loss of organic carbon (kg/ha)**
0.5	6.6
1.5	13.9
3.0	46.0
Slope length	
18.3	13.8
36.6	7.1
54.9	4.9

Source: (18)

The intensive cultivation stimulates decomposition of soil organic matter (SOM). Organic carbon status usually remains low in cultivated soils. It is clear that in all the soil zones, the organic matter content is very high in virgin soil (Table 3.10).

Table 3.10 Organic matter content (%) in virgin and cultivated soils

Soil zone	Virgin	Cultivated
Brown	3–4	2–3
Dark brown	4–5	3–4
Black	6–10	4–6
Dark grey	4–5	2–3

Source: (18)

The total biomass (t/ha) accumulated at the time of observation was maximum in natural forest *S. robusta (NFSR)*, followed by plantation of *D. sissoo* (PDS), tea garden (TG), plantation of *T. arjuna* (PTA), mango-based agri-horticulture agroforestry system (AHAF), agricultural field (AF) and fallow land (FL) (Fig. 3.4). However, while considering the rate of accumulation of biomass per unit area and per unit time with reference to the total biomass accumulated, land use systems based on annual crops exceeded the values compared to those having perennials as dominant components (Fig. 3.5). Taking the amount of biomass accumulated per year, natural forest had an edge over the rest of land uses followed by PDS, TG, PTA and least in FL. Converting biomass into carbon stored into all the land uses followed by PDS, TG, PTA and least in FL. The estimate of biomass accumulation under different land uses and their corresponding carbon stock are given in Figs 3.4 and 3.5.

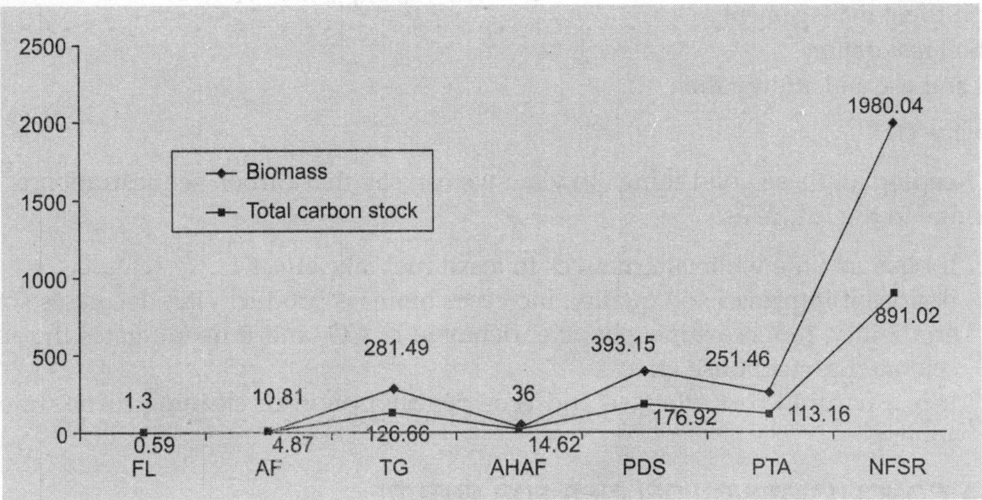

Fig. 3.4 Biomass and carbon stock (t/ha) in different land use

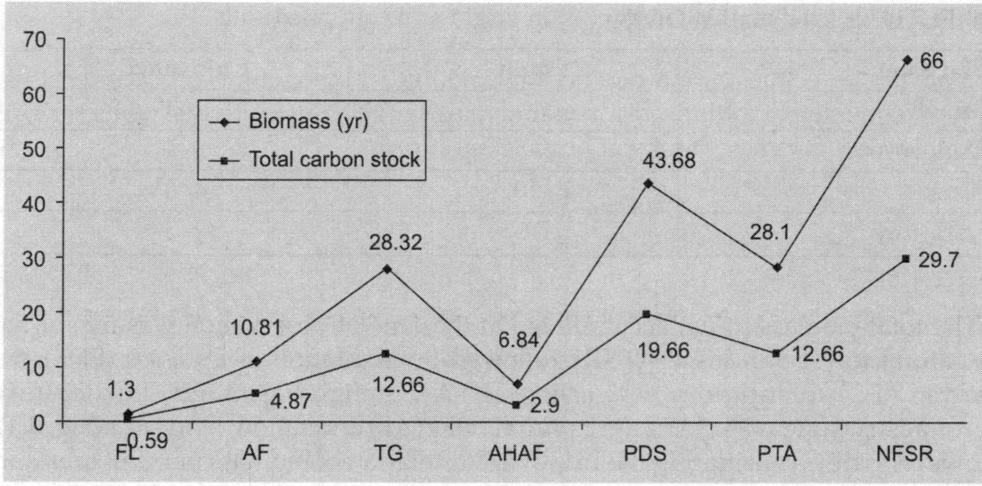

Fig. 3.5 Biomass and carbon increment (t/ha/yr) under different land uses

The conservation tillage, cover crops, INM, soil restoration, land use, etc. if implicated together in a proper way can contribute to the additional carbon sequestration of 100–1000 kg C/ha/yr.

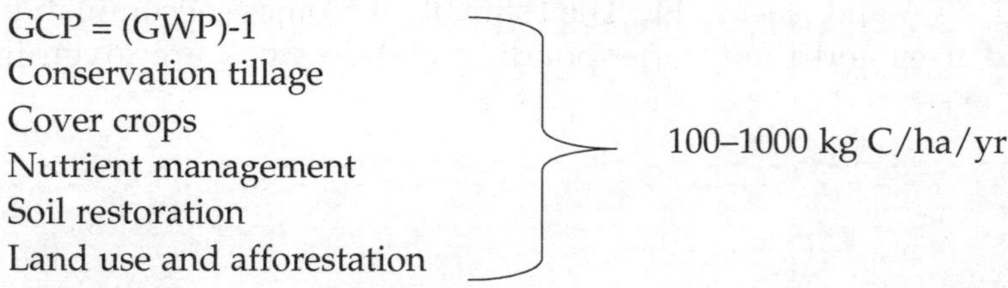

Source: (12)

Keeping all these good things in view we can say that carbon sequestration is a bridge to the future as:

1. It buys us time while alternatives to fossil fuel take effect.
2. Besides it improves soil quality, increases biomass productivity, decreases soil erodibility, reduces atmospheric enrichment of CO_2 and thus mitigates the accelerated green house effect.
3. It is a natural, cost effective and eco-friendly option of cleaning dirty atmosphere.

3.4 CONCLUSIONS/GUIDELINES FOR FUTURE

The SOC is a key indicator of soil health for sustainability and productivity of soils and crop management. Soil organic carbon is an extremely valuable natural resource and irrespective of the climate debate, the SOC stock must be restored, enhanced and improved. Conserving technologies coupled INM practices are important for restoring SOC in cultivable lands.

3.5 FUTURE LINE OF WORK

- Soil carbon sequestration practices (SCSPs) must be documented area and region wise and popularised for maintenance of SOC in cultivable soils for soil quality enhancement and soil health restoration.
- There should be a paradigm shift in LUPs from resource degrading to resource conserving technologies (RCTs) for possible areas of high intensive agriculture.
- There is a need for a commission for safer use of soils and for soil, water and plant analysis protocols.
- The awareness should be created about soil, plant and water resources through a series of *Kisan Melas*, farmer's camps, trainings, workshop, symposiums, conferences.
- Land use management practices for soil carbon sequestrations and farming carbon for mitigating global climate change must be harmonised and popularised for benefits of farmers besides improving soil fertility and land productivity.
- A computer added database on adoption of recommended agricultural practices for enhancing soil organic carbon and improving soil fertility crops, regions and state wise for better management of soils, crops, water and plant nutrient sources for future agriculture.

3.6 KEY REFERENCES AND RESOURCES FOR FURTHER READING

1. Balesdent JC, Chenu, Balabane M. Soil and Tillage Res 2000; 53:215–220.
2. Berdsey RA. Forest and Global Change 1996; 2:1–25.
3. Bhojvaid PP, Timmer VR. Forest Ecol. Management 1998; 106:181–193.
4. Brar MS, Sharma Preeti, Indian J Fert 3:27–30.
5. Chand S. In Integrated Nutrient Management for Sustaining Crop Productivity and Soil Health. IBD Co, Lucknow 2008; p112.
6. Divy NK, Panwar P. Current Sci. 2008; 95:5:658–663
7. Gupta RK, Rao D LN. Current Sci. 1994; 66:378–380.
8. Houghton RA. Soils and Global Change 2003; 45–65.
9. IPCC, Land Use, Land Use Change, and Forestry. A Special Report of IPCC. Cambridge Univ. Press, UK 2000; 377.
10. IPCC, A Synthesis Report. Cambridge Univ. Press, UK 2001; p398.
11. IPCC, A Synthesis Report. Cambridge Univ. Press, UK 2007; p2.1.
12. Lal R. Soil and Tillage Res 1997; 43:81–107.
13. Lal R. Prog Env Sci 1999; 1:307–326.
14. Lal R. Global Climate Change and Tropical Ecosystems. CRC Press 2000.
15. Lal R. Critical Reviews in Plant Sci 2003; 22:151184.
16. Nambiar KKM. Agricultural Sustainability — Economic, Environment and Statistical Considerations. 1995, J Wiley and Sons, New York.
17. Schuman GE, Janzen HH, Xerrick JE. Environ. Pollution 2002; 116:391–396.

18. Stan DW, Steve AS, Donald LR, Charles TG. Environ. Sci. 2000; 30:2091–2098.
19. Swarp A, Reddy DD, Prasad RN. In Proc. of a Natural Workshop on Long Term Soil Fertility Management Through Integrated Plant Nutrient Supply. IISS, Bhopal 1998.
20. Vandana V, Sheela KR. Indian J Fert 2008; 10:45–49.
21. West TO, Post WM. Soil Sci Soc of America J 2002; 66:1930–1936.
22. Xiubin L, QiGuo Z. Climate Change 2002; 40:119–133.
23. Chand S, Singh L, and Singh P. Sustainaible agriculture, food security and climate change. Daya Publishing House, N. Delhi, 2012; p218.
24. Chand S. (2010) Challenges of Soil Quality of Indian Soils *vis-à-vis* Food Security. Current Science, Indian Academy of Sciences, Bangaluru.

CHAPTER 4 Farming Carbon and Sequestration for Enhancing Soil Organic Carbon (SOC) in Agricultural and Cultivable Lands

Have u ever wanted to be a superhero! You can be a hero to the 1 billion kids that go to bed hungry every night. — World Food Program

In the past four decades, due to heavy exploitation of agricultural lands through intensive cropping, degradation and indiscriminate use of agricultural inputs like chemical fertilizers, pesticides, herbicides results in deterioration of soil quality (physical, chemical and biological properties). Mono cropping again doubled exploitation of lands, results in low soil organic carbon. Soil organic carbon presence in agriculture lands regarded as universal soil quality indicator, had major impacts on productivity of agriculture lands. Modern research and technologies for improving agriculture lands shrinking towards increasing soil organic carbon through carbon farming and sequestration. Farming carbon not only improves productivity of agriculture lands, but helps in mitigation of climate change impact on agriculture. This review highlights the various methods of farming carbon and sequestration in agriculture lands for sustainable crop production and soil health.

4.1 PREAMBLE

The earth's atmosphere contains carbon dioxide (CO_2) and other greenhouse gases (GHGs) that act as a protective layer, causing the planet to be warmer. This heat retention is critical to maintaining habitable temperatures. If there was significantly less CO_2 in the atmosphere, global temperatures would drop below levels to which ecosystems and human societies have adopted. As levels rise, mean global temperatures are also expected to rise as increasing amounts of solar radiation are trapped inside the 'greenhouse'. The concentration in the atmosphere is determined by a continuous flow among the stores of carbon in the atmosphere, the ocean, the earth's biological systems, and its geological materials (Fig. 4.1). As long as the amount of carbon flowing into the atmosphere (as CO_2) and out (in the farm of plant material and dissolved carbon) are in balance, the level of carbon in the atmosphere remains constant. According to IPCC, the three main causes of the increase in GHGs observed over the past 250 years have been fossil fuels, land use, and agriculture. Agriculture is itself responsible for an estimated one third of global warming and climate change. It is generally agreed that about 25 per cent of the main GHGs, CO_2, is produced by agricultural sources, mainly deforestation and the burning of biomass.

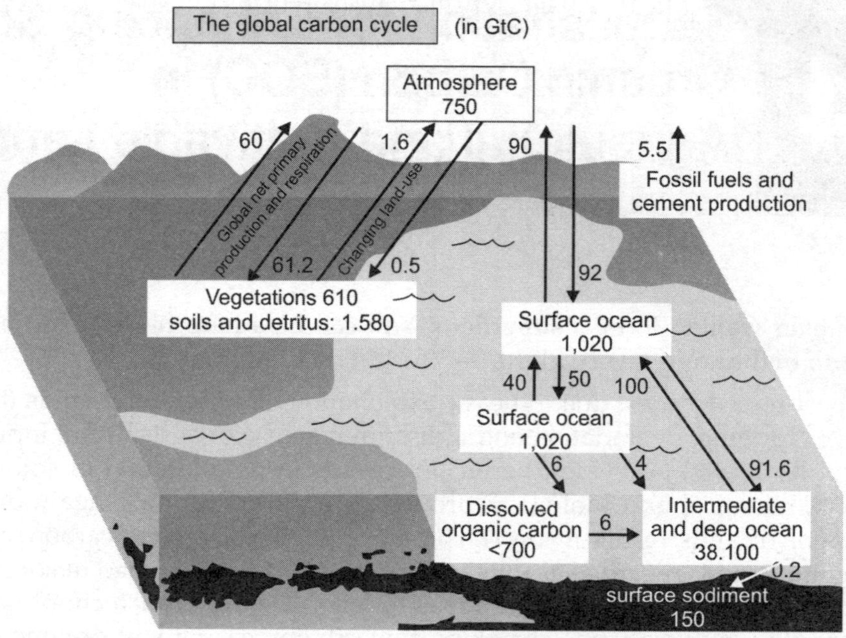

Fig. 4.1 Global carbon cycle [Source: www.met office.gov.uk/research/hadleycentre/models/carbon_cycle/intro_global]

Figure 4.1 illustrates that the primary sources of the slow, but steady increase in atmospheric carbon that is now occurring are fossil fuel combustion, which contributes approximately 5.5 gigatons (billion metric tons) of carbon per year, and land use changes, which account for another 1.1 gigatons. In contrast, the oceans absorb from the atmosphere approximately 2 more gigatons of carbon than they release and the earth's ecosystems appear to be accumulating another 1.2 gigatones annually. In all, the atmosphere is annually absorbing approximately 3.4 gigatones of carbon more than it is releasing.

While the annual net increase in atmospheric carbon does not sound large compared with the total amount of carbon stored in the atmosphere —750 gigatones — it adds up over time. For example, if the current rate of atmospheric carbon accumulation were to remain constant, there would be a net gain in atmospheric carbon levels of 25 per cent over fifty years. In fact, the rate at which human activity contributes to increases in atmospheric carbon is accelerating. Table 4.1, describes trends in world carbon emissions from fossil fuel combustion and land use change since 1850. There has been a steady increase in carbon emission from fossil fuel use at global levels. Emission from land use change have also been growing at the global level, though not nearly as rapidly as from fossil fuel combustion.

Table 4.1 World and the US emissions from fossil fuels and land-use change selected years (millions of metric tons of carbon per year)

	1850	1900	1950	1960	1970	1980	1990	2000
Fossil fuel								
World	54	534	1612	2535	3908	5177	5959	6395
Land-use change								
World	503	607	935	1302	1537	1608	2158	2081

Source : Land-use Change data from "Annual Net Flux of Carbon to the Atmosphere from Land-Use-Change: 1850-2000: at http://cdiac.ornl.gov/ftp/trends/landuse/houghton/houghtondata.txt.
Fossil fuel emission data from "Global CO_2 emissions from Fossil-Fuel-Burning, Cement Manufacture, and Gas Flaring: 1751-2000" at http://cdiac.ornl.gov/ftp/ndp030/globa100.ems, and "Global CO_2 Emissions from Fossil-Fuel Burning, Cement Manufacture, and Gas Flaring: 1751-2000" at http://cdiac.ornl.gov/ftp/ndp030/nation00.ems.

Climate change and agriculture are interrelated processes, both of which take place on global scale. Global warming is projected to have significant impacts on conditions affecting agriculture including temperature, precipitation and glacial runoff. These conditions determine the carrying capacity of the biosphere to produce enough food for the human population and domesticated animals. Agriculture besides being major source of GHGs, can also be a sink for carbon in the form of soil organic matter. Crops absorb CO_2 from the atmosphere and use the carbon to produce organic matter is a process called carbon fixation. Thus, changes in agricultural management practices can induce dramatic changes in soil organic carbon and as a result a new agriculture has been designed for the era of climate change called carbon farming — the farm plan which incorporates both carbon capture and emission reduction.

4.2 FARMING CARBON (FC)

Farming carbon is a new way to describe a collection of eco-friendly farming technique which increases soil organic carbon pools in agricultural lands. These long lived pools can be living, above ground biomass (e.g. trees), products with a long, useful life created from biomass (e.g. lumber), living biomass in soil (e.g. roots and microorganisms), or recalcitrant organic and inorganic carbon in soils and deeper subsurface environments. The various methods of carbon farming for capturing and holding carbon in soil includes, establishment of permanent vegetative cover as in the conservation reserve programme (pasture management)[1,2]; conservation tillage practices such as no-tillage[3,4]; increased return of organic to soil through perennial crops and greater yields of annual crops and reduction of fallow periods[5,6]; grazing management; biological farming; mulching and afforestation. Soil management practices that usually improve soil organic matter include; *i.* more complex crop rotation, especially those with high residue crops, *ii.* intensive use of cover crops, *iii.* use of variety of organic amendments, *iv.* balanced fertilization, and reduced tillage.

The role of agricultural soil in the production and consumption of GHGs of the more recent scientific interest[7-10] while current contributions of land use to atmospheric CO_2 increases (approximately 25%)[11] are mainly from tropical regions, the rise in CO_2 during the 19th and early 20th centuries was largely associated with the rapid expansion of temperate zone agriculture. As much as 110 Pg (10^{15} gm) of C may have been released from temperate zone ecosystems, contributing substantially to increase CO_2 prior to the mid-1900s[12]. Currently, the C balance in temperate agricultural soils appears to have stabilized[13], but there is considerable interest in evaluating the potential of these soils to regain some of their lost C and help ameliorate the continued increase in atmospheric CO_2[13, 14]. There is a tremendous store of empirical knowledge about the influence of management of SOM, and the practices which promote SOM formation and maintenance are well known. These practices include:

4.2.1 Low/No Tillage/Minimum Tillage/Zero Tillage

The idea of tillage is fundamental to most people's view of agriculture. Jethro Tull energetically promoted the benefits of intensive tillage as a means of breaking down the soil into minute particles which could then be absorbed by plants[15]. Although his theory of plant nutrition was subsequently refuted, Tull's prescription contained an element of truth in that part of the effectiveness of intensive tillage is to speed the release of nutrients contained in SOM. However, over the long term, intensive tillage has caused or contributed to soil degradation in many regions. Reducing erosion and organic matter losses in cultivated soils has been the primary reason for the development of less intensive tillage practices[16-18]. In conventional tillage three or four tillage operations are followed so as to make the soil more friable to be congenial for germination of seeds at the cost of easily decomposable SOM. In order to preserve organic carbon in soil, the conservation tillage practices like no tillage, zero tillage and ridge tillage has been gaining popularity. These come under the umbrella of conservation tillage. By minimizing soil tillage and its associated emissions, global increase of atmospheric CO_2 can be reduced while at the same time increasing soil carbon deposits (sequestration) and enhancing soil quality.

The IPCC (2006) greenhouse gas inventory guidelines suggested that conversion from conventional tillage to no tillage systems leads to a 10 per cent increase in the estimated sequestration of carbon in the soil. Based on the data from a number of long term field studies, with paired conventional tillage (CT) and no tillage (NT) treatments, the average increase in soil C under NT was about 300 gm^{-2} with a few site/treatments showing increase as high as 1 $kgcm^{-2}$. On a relative basis, most sites showed 5 to 20 per cent increase in soil C under NT vs CT[19, 20]. In 25-year-old tillage experiments, Dick *et al.*[21] reported that the most rapid changes in C levels under NT occurred during the first 10 years. Repeated sampling of NT plots over a 13 year period showed an increase of 15 per cent in soil C after 6 years, whereas C increased by only an additional 4 per cent during the subsequent 7 years.

Studies have shown no or variable differences in crop yield response to tillage[22–24] and higher production under no till has been shown for well drained soil, particularly where water use efficiency is improved[17, 25]. Prasad and Power[26] summarized the expected differences in yield between NT and CT as follows: *i.* little difference under conditions of adequate soil water, good drainage and adequate available N, *ii.* increased yields under NT where there is limited precipitation and soil water and adequate weed control and fertilization, and iii. reduced yields under NT in areas with excessive precipitation, low temperature, poor drainage, poor weed control or low fertility levels. To summarize, it seems clear that reduced tillage and especially NT, is generally effective in increasing soil C, provided that yields and residue production are not adversely affected.

4.2.2 Crop Rotations/Cropping Sequence

Compared with monoculture cropping practices, multi crop rotations with two or three crops in a year can result in increased SOC contents (Table 4.2). This is because of addition of large amount of above ground as well as underground biomass in soil. Some crops grown in rotation leaves especially large quantities of residue, contributing greatly to the addition side of the gains-losses equation. Some crops, such as legumes, grasses, or grass legume forage crops, supply a lot of root biomass, which can contribute to residue. As compared to tomato inclusion of sunflower in pearl millt-potato cropping system enhanced SOC contents. This is further heightened due to introduction of leguminous green manuring crops like dhaincha (*Sesbania aculeate* L.) in pearlmillet-wheat-dhaincha[27], cowpea in sorghum cowpea[28], and clover in rice clover crop rotations[29].

Table 4.2 Effect of crop rotation on soil organic carbon contents in soils of sub-humid, semi- and tropical ecosystems

Cropping system	**Soil organic carbon (g ka^{-1})**	**Microbial biomass carbon (mg kg^{-1})**	**Total nitrogen (%) available nitrogen (kg ha^{-1})**	**Reference**
	Inceptisol, Hisar, Haryana, 6 years			
Pearmillet-Wheat-fallow	4.83	192	0.060#	27
Pearmillet-Fodder-cowpea-fallow	4.78	231	0.061	
Pearmillet-Potato-Tomato	5.04	221	0.063	
Pearmillet-Potato-Sunflower	4.54	184	0.059	
Pearmillet-Mustard-Sunflower	5.02	144	0.062	
LSD (p=0.05)	0.09	11	0.005	

Contd.

Table 4.2 Effect of crop rotation on soil organic carbon contents in soils of sub-humid, semi- and tropical ecosystems *(Contd.)*

Cropping system	Soil organic carbon (g ka^{-1})	Microbial biomass carbon (mg kg^{-1})	Total nitrogen (%) available nitrogen (kg ha^{-1})	Reference
	Vertisol, Hisar, Haryana, 2 years			
Sole Sorghum	6.21		230*	28
Sorghum-cowpea	7.07		233	
	Inceptisol, Hisar, Haryana, 5 years			
Rice-mustard	5.2	201.7	0.064#	29
Rice-wheat	6.0	270.0	0.055	
Rice-clover	7.2	241.8	0.083	
Sorghum-wheat	5.5	231.2	0.0061	
CD (p=0.05)	1.4	NS	0.001	

total N in per cent, * available N in kg ha^{-1}, CD critical difference, LSD: = least significant difference, NS = not significant

4.2.3 Plant Nutrient Applications

Nitrogen availability can influence soil C levels in a variety of ways. It is clear that by increasing crop production, and thereby residue inputs, N fertilization can contribute to increased SOM contents. By increasing plant growth, fertilization can also lead to increased transpiration, drier soils and decreased decomposition rates[30]. Results from many long term studies show a general tendency of increases in soil C with substantive additions of N, compared to zero or low N additions. Nitrogen additions can affect decomposition rates and C stabilization efficiency in other ways that contribute to higher SOM levels. Fog[31] reviewed 60 papers which reported zero or negative effects of N addition on decomposition rates. He offered several possible explanations for negative effects of N additions on decomposition, including the repression of lignolytic enzymes by ammonium and an increase in the amount of amino compounds which can act as precursors in the formation of recalcitrant humic compounds. At the microbial level, insufficient N can lead to lower yield efficiencies (i.e. more CO_2 respired per unit C assimilated[32]). Under such conditions addition of N could increase growth efficiency resulting in a higher proportion of C inputs retained in SOM.

In an analysis of long term plot studies with constant above ground C, with and without N fertilization, Paustain *et al.*[33] found that increased root residue inputs could not account for observed increases in soil C in the N fertilized treatments (Fig. 4.2). They also found that soil C — N ratio were lower in the unfertilized treatments, where straw or sawdust were added, suggesting that N limitation may have reduced C stabilization efficiency. In a study by Campbell *et al.*[34], soil C was found to be similar in fertilized plots where straw was removed compared with fertilized plots where straw was retrained. This was despite the fact that C inputs were estimated to be greater in the treatment with straw retention.

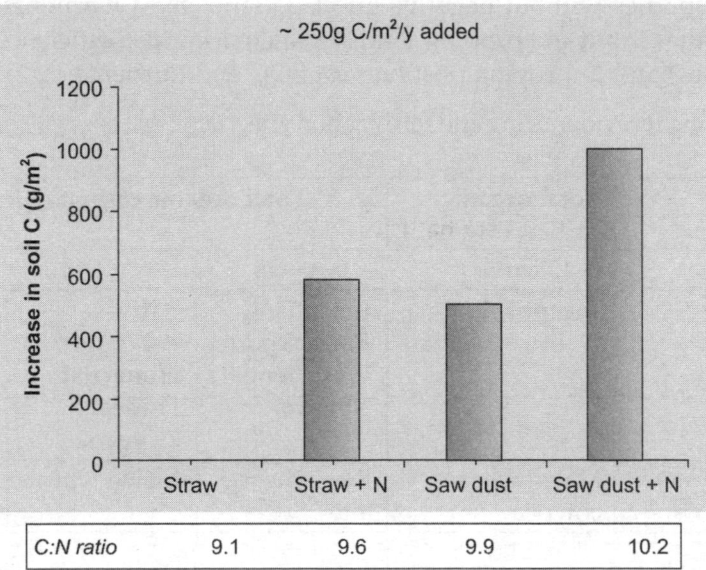

Fig. 4.2 N addition effect on the net change in soil C over the duration of a 31 years experiment with constant (~250 g cm^{-2} y^{-1}) of straw or sawdust addition above ground crop residues were removed from the plots. N fertilized plots received rates equivalent to 80 kg N ha^{-1} y^{-1} as $Ca(NO_3)_2$.

In most field experiments it is difficult to partition the interacting and potentially conflicting, effects of N addition as soil C. However, when viewed in a broad sense, it is reasonable that, since C and N are the major constituents of SOM and their proportionality (i.e. C—N ratio) is relatively constant across a range of agricultural soils, then an adequate supply of N is required to build SOM. If inputs of these two elements are too much out of balance then the efficiency of soil C sequestration will be reduced. The balanced application of NPK (100% or 150% NPK) also showed higher accumulation of SOC over imbalanced use of fertilizers (100% N and 100% NP) in different cropping systems (maize-wheat-cowpea, rice-wheat-jute, maize-wheat, and soybean-wheat) over three decades under dissimilar climate and soil (Table 4.3)[35, 36, 37, 38]. Total organic carbon content in entire 0–45 cm soil profile in maize-wheat-cowpea cropping system followed the order: 150% NPK + FYM > 150% NPK > 100% NPK > 100% NP = 100% N = 50% NPK > control.

Chand[53] complied 200 applied integrated nutrients management (INM) practices for crops and cropping systems of various states of the country, found that integration of nutrient sources both organic and inorganic results in sustainable yield and enriches SOC of soils without deterioration in soil texture and fertility. Similarly, Chand[54] in an experiment on integrated nutrient management in mustard conducted at Udaipur, found that 50 per cent of recommended dose of NPK's coupled with FYM @ 10 t ha^{-1} + use of microbial inoculants (*Azotobactor* + PSB) helps in enriching yield of mustard besides improving soil physical, biological and chemical properties without negative effect on land. INM harmonizes soil, crop, water and other input in cropping lands by increasing use efficiencies (fertilizer, water or input) and improving post harvest SOC and nutrients.

Table 4.3 Long term manuring and fertilization effects on total organic carbon contents in soil

Treatment	**Total organic carbon (Mg ha^{-1})**	**Total organic carbon (g kg^{-1})**		
	0–45 cm	**0–15 cm**		**0–45 cm**
	Inceptisol, Delhi, 31 years	**Inceptisol, Barrackpore, West Bengal, 30 years**	**Alfisol, Palampur, Himachal Pradesh, 29 years**	**Inceptisol, Almora, Uttarachal, 30 years**
Cropping system	Maize-wheat-cowpea	Rice-wheat-jute	Maize-wheat	Soybean-wheat
50% NPK	51.5	–	9.4	–
100% NPK	54.1	7.4	8.9	22.81
150% NPK	53.5	–	–	–
100% NP	53.0	6.3	7.3	21.05
100% N	52.1	5.7	8.6	–
100% NPK + FYM	72.1	7.9	12.7	31.42
Control	48.7	5.1	7.1	18.08
CD (p=0.05)	–	–	0.4	–
Reference	37	36	38	35

The balanced fertilization with major, secondary and micronutrients in the key to enhance organic carbon contents in soil. Better crop growth in terms of both above and underground by balanced fertilization was the main reason for this. Application of major nutrients in right proportion ($N_{120}P_{60}K_{60}$) increased organic carbon than imbalanced fertilizer treatments. However, when FYM was superimposed over 100% N there was significant build up of total organic carbon contents (Table 4.4)[38, 39]. Not only the major nutrients, applications of secondary and micronutrients through $ZnSO_4$ with 100% NPK showed higher levels of organic carbon as those of the treatments FYM @ 5 t ha^{-1} or groundnut shells @ 4 t ha^{-1}[40]. Even green manure is reported to be superior to FYM increasing organic carbon in soil. As compared to FYM, compost is reported to be more efficient in enhancing organic carbon contents in soil[41]. As an alternative source, press mud cake not only economized on nitrogen, but also benefitted the soil by increasing carbon content[42].

Table 4.4 Effect of organic, inorganic and their combined applications on total organic carbon, available nutrients, microbial biomass and yield of crops in different soils of India

Treatment	Total organic carbon ($g\ kg^{-1}$)	Available nutrients N	P	K	S	Fe	Mn	Zn	Cu	Microbial biomass C ($mg\ kg^{-1}$)	Yield ($t\ ha^{-1}$)	Reference
		($kg\ ha^{-1}$)				($mg\ kg^{-1}$)						
	Inceptisol, Jammu & Kashmir, 6 years, cropping system: Rice-wheat											
100% N	3.8	262	10.5	90.3	15.8	15.8	11.8	0.70	0.85		3.11 (wheat)	39
100% N + FYM @ 5 t ha^{-1}	5.5	362	17.6	108.2	22.6	20.8	15.2	1.04	1.22		3.44	
CD (p=0.05)	1.2	6.5	1.3	4.9	1.6	2.3	1.2	0.15	0.22		0.12	
	Alfisol, Anantapur, Andhra Pradesh, 17 years, cropping system: Groundnut											
100% NPK	2.3	110.0	54.0	140.0								40
100% NPK + $ZnSO_4$	2.6	102.0	58.0	108.0								
FYM @ 5 t ha^{-1}	2.6	120.0	18.0	188.0								
Groundnut shells @ 4 t ha^{-1}	2.7	94.0	25.0	101.0								
Control	1.9	112.0	14.0	103.0								
CD (p=0.05)	0.09	NS	8.4	43.0								
	Vertisol, Bijapur, Karnataka, 2 years, cropping system: Sorghum-chickpea											
50% N FYM + 50%	4.7	220.0	21.8	451.0								43
Recommended dose of N (RDN)	5.4	221.0	25.6	403.0								
50% N compost + 50% RDN												
50% N vermicompost + 50% RDN	4.7	218.0	25.6	405.0								
100% NPK	5.7	180.0	25.1	37.0								
Control	4.5	150.0	203	40.1								
CD (p=0.05)	0.02	1.6	0.8	0.75								

Contd.

Table 4.4 Effect of organic, inorganic and their combined applications on total organic carbon, available nutrients, microbial biomass and yield of crops in different soils of India *(Contd.)*

Treatment	Total organic carbon ($g\ kg^{-1}$)	Available nutrients								Microbial biomass C ($mg\ kg^{-1}$)	Yield ($t\ ha^{-1}$)	Reference
		N	P	K	S	Fe	Mn	Zn	Cu			
		($kg\ ha^{-1}$)				($mg\ kg^{-1}$)						
	Inceptisol, Almora, Uttaranchal, 3 years, Cropping system: Rice-wheat											
240 kg N ha^{-1} + 60 kg P ha^{-1}	3.2		4.9									42
Press mud cake (PMC) + 140 kg N ha^{-1} + 30 kg P ha^{-1}	3.9		12.1									
PMC + 180 kg N ha^{-1} + 60 kg P ha^{-1}	4.0		12.3									
	Inceptisol, Karnal, Haryana, 4 year, Cropping system: Rice-wheat											44
$N_{120}\ P_{25}\ K_{42}$	2.9	178.0	14.1	225.0								
$N_{120}\ P_{26}\ K_{42}$ + FYM	2.9	193.0	19.0	262.0								
$N_{120}\ P_{26}\ K_{42}$ + Green manure	3.9	195.0	19.0	285.0								
Control	4.2	101.0	9.3	187.0								
CD (p=0.05)	0.5	11.0	1.3	16.0								

4.2.4 Land Use Management Practices (LUMPs)

Vermani[45] noted that according to some environmentalists, we need to reduce our agricultural area from the current 140 million ha to 110 million ha by the year 2000 to 2010. The area thus released can be used for alternate land use, most likely forestry because of the low proportion of land area under forest cover (11%) against the required (23%). Alternate land use systems, viz. agro-forestry, agro-horticulture and agro-silviculture are more remunerative for SOC restoration as compared to sole cropping system[41]. Das and Intal[46] reported that organic carbon content was about double in agro-horticulture and agro-forestry systems as compared to sole cropping (Table 5.5). But Purkayastha *et al.*[47] reported that vegetable growing plots exhibited similar SOC contents as rice-wheat growing plots. Soil organic carbon and microbial biomass carbon in agro-forestry was significantly higher than in either of the two systems reported above. Dhaliwal[48] showed higher SOC and microbial biomass carbon in natural forest system followed by cultivated and pasture ecosystem.

Table 5.5 Effect of various land uses on soil organic carbon and microbial biomass carbon[48]

Various land uses	Soil organic carbon	Microbial biomass carbon	Soil organic carbon			Microbial biomass carbon
	(Mg ha^{-1}), 0–20 cm		(g kg^{-1}), 0–15 cm			
Field cropping	21.1	0.48	4.2	12.3	4.8	100
Agro-forestry	33.7	0.80	7.1	–	–	–
Agro-horticulture	20.4	0.54	7.3	–	–	–
Agro-siliviculture	–	–	3.8	–	–	–
Wasteland	–	–	–	19.2	–	–
Forest	–	–	–	20.4	5.6	125
Grassland	–	–	–	16.2	4.2	94
Orchard	–	–	–	15.8	–	–
Hops	–	–	–	12.4	–	–
References	47	46		52		48

4.2.5 Biochar Burial

Biochar is charcoal created by pyrolysis of biomass. The type of carbon contained in biochar is black carbon. The resulting charcoal like material is land filled or used as a soil improver to create terra preta highly fertile soils rich in black carbon, i.e. type of carbon found in charcoal. Thirteen governments as well as the United Nations convention to combat desertification (UNCCD) are formally calling for 'biochar' to play a significant role in a post 2012 climate change agreement and in carbon trading. They support claims by the international biochar initiative (IBI).

The IBI argues that applying charcoal to soil creates a reliable and permanent carbon sink and mitigates climate change as well as making soils more fertile and water retentive. However, even the studies of IBI members and supporters indicate high levels of uncertainty and counter-indications. In addition, proponents of biochar do not consider the direct and indirect impact of land-use changes required to grow enough biomass raw materials, or the impact of removing large quantities of so called residues from fields and forests that would occur with large scale production systems. The quantities of biochar proposed for mitigating climate change would require more than 500 million hectares of additional plantation[49]. Johannes Lehmann[50] reported that only 1 to 20 per cent of the carbon in charcoal will be lost in the short term and that the remainder will stay in the soil for thousands of years. Yet, one study about the fate of black carbon from vegetation burning in Western Kenya suggests that 72 per cent of the carbon was lost within 20 to 30 years[51].

4.2.6 Composting Technologies

Composting is an important process by which farm biomass, crop residues can be converted into a usable form by a certain process like phospho-composting, vermi-composting and enriched-compositing. It helps in converting unstable solids into stable carbon solid compounds restricted further changes in organic carbon in soil. Composting uses in agriculture have now encouraged in every state for improving soil physical properties through enhancement in soil carbon of agricultural lands. It is an important component of INM practices. Vermin-compost is found superior over all composts for enriching SOC of agriculture lands[54].

4.3 CONCLUSIONS

Global climate change adversely affects systems or sectors. The SOC of agricultural lands continuously decreases due to several causes including climate change, loss of SOC by intensive cropping without adding carbonic biomass to agricultural lands. Farming carbon, fertilization through organics and inorganic, restricted tillage practices, introduction of crop rotations, composting and biochar are important methods of enriching SOC and mitigating global climate change impact on agriculture by reducing GHGs emissions. Sequestration of carbon through improved agriculture practices is imperative for soil health improvement for sustainable crop production. Improving SOC content of agricultural lands will results in good crop stand and may decrease further degradation of lands, since SOC is a unique indicator of productivity and fertility of arable land. Farming carbon may be increased through composting (converting crop residues and biomass into a usable form at farm sites by phospho-composting, enriched-composting and vermicomposting). Farming carbon practices needs to be identified and popularized for the benefits of agricultural farming communities at farm and village level for improving socio-economic status and sustainable agricultural development of Indian farming communities without further degradation of agricultural lands for future uses. Our soils are basis for even society development should not be abused further. In the changing modern agriculture industrial scenario, it is imperative to use climate and soil friendly approaches and cultural practices for maintaining soil health on long term basis for future generations and to overcome climate change impact on agriculture.

4.4 KEY REFERENCES AND RESOURCES FOR FURTHER READING

1. Huggins DL, Allan DR, Gardner JC, *et al.* In: R Lal *et al.* (ed.) Adv Soil Sci Lewis-CRC Press, Boca Raton, FL 1997; p323–334.
2. Ogle SM, Breidt FJ, Eve MD, *et al. Global Change Biol* 2003; 9:1521–1542.
3. Bhattacharya R, Prakash V, Kundu S, *et al. Indian Soc Soil Sci* 2004; 52:238–242.
4. Halvorson AD, Weinhold BJ, Black AL. *Soil Sci Soc Am J* 2002; 66:906–912.
5. Huggins DR, Buyanovsky GA, Wagner GH, *et al. Soil Tillage Res* 1998; 47:219–234.
6. Machado S, Rhinhart K, Petrie S. *J Environ Qual* 2006; 35:1548–1553.
7. Bouwman AF (Ed.) Soils and the Greenhouse Effect. John Wiley & Sons 1990; Chichester, England.
8. Duxbury JM, Harper LA, Mosier AR. Contributions of Agro-Ecosystems to Global Climate Change. In: Agricultural Ecosystem Effects on Trace Gases and Global Climate Change. ASA Special Publication 1993; 55:1.
9. Wisniewski J, Lugo AE (Eds). Natural Sinks of CO_2. Kluwer Academic 1992; Dordrecht.
10. Wisniewski J, and Sampson RN, (Eds). Terrestial Biospheric Carbon Fluxes: Quantification of Sinks and Sources of CO_2 Kluwer Academic 1993; Dordrecht.
11. Houghton RA, Skole DL. Carbon, in The Earth as Transformed by Human Action BL, Turner WC, Clark RW, Kates, *et al.* (Eds) Cambridge University Press, Cambridge 1990; Chap. 23.
12. Wilson AT. Pioneer Agriculture Explosion and CO_2 Levels in the Atmosphere. Nature 1978; p273–240.
13. Cole CV, Paustain K, Elliott ET, Analysis of Agroecosystem Carbon Pools. Water, Air, Soil Pollut 1993a; 70:357.
14. Cole CV, Flach K, Lee J, Agricultural Sources and Sinks of Carbon. Water, Air, Soil Pollut 1993b; 70:111.
15. Fussel GE. Farming Technique from Prehistoric to Modern Times. Pergamon Press 1965; Oxford.
16. Baeumer K and Bakermans WAP. Zero Tillage. *Adv Agron* 1973; 25:77.
17. Phillips RE, Blevins RL, Thomas GW, *et al.* No Tillage Agriculture. *Science* 1980; 208:1108.
18. Gebhardt MR, Daniel TC, Schweizer EE, *et al.* Conservation Tillage. *Science* 1985; 230:625.
19. Halvin JL, Kissel DE, Maddux LD, *et al.* Crop Rotation and Tllage Effects on Soil Organic Carbon and Nitrogen. *Soil Science Soc Am J* 1990; 54:448.
20. Ismail I, Blevins RL, Frye WW, Long Term No Tillage Effects on Soil Properties and Continuous Corn Yields. *Soil Sci Soc Am J* 1994; 58:193.

21. Dick WA, McCoy EL, Edwards MW. Continuous Application of No Tillage to Ohio Soils. *Agron J* 1991; 83:65.
22. Francis GS, Cameron KC, Swift RS. Soil Physical Conditions After Six Years of Direct Drilling or Conventional Cultivation on a Silt Loam Soil in New Zealand. *Aust J Soil Res* 1987; 25:517.
23. Coote DR and Malcolm-McGovern CA. Effects of Conventional and no Toll Corn Grown in Rotation on Three Soils in Eastern Ontario, Canada. *Soil Tillage Res* 1989; 14:67.
24. White PF. The Influence of Alternative Tillage Systems on the Distribution of Nutrients and Organic Carbon in Some Common Western Australian Wheat Belt Soils. *Aust J Soil Res* 1990; 28:95.
25. Carefoot JM, Nyborg M, Lindwall CW. Tillage-Induced Soil Changes and Related Grain Yield in Semi arid Region. *Can J Soil Sci* 1990; 70:203.
26. Prasad R, Power JF. Crop Residue Management. *Adv Soil Sci* 1991; 15:205.
27. Chander K, Goyal S, Mundra MC. *Biol Fertil Soils* 1997; 24:306-310.
28. Duriasami VP, Perumal R and Mani AK. *J Indian Soc Soil Sci* 2001; 49:435–439.
29. Batra L, Rao DLN 2004; Under publication.
30. Andren O. Decomposition in the Field of Shoots and Roots of Barley, Lucern and Meadow Fescue. *Swed J Agric Res* 1987; 17:113.
31. Fog K. The Effect of Added Nitrogen on the Rate of Decomposition of Organic Matter. *Biol Rev* 1988; 63:433.
32. Tempest DW, Neijssd OM. The Status of Y_{ATP} and Maintenance Energy as Biologically Interpretable Phenomenon. *Ann Rev Microbiol* 1984; 38:459.
33. Paustain K, Parton WJ, Persson J. Medeling Soil Organic Matter in Organic Amended and Nitrogen Fertilized Long Term Plots. *Soil Sci Soc Am J* 1992; 58:476.
34. Campbell CA, Lafond GP, Zentner RP, *et al*. Influence of Fertilizer and Straw Baling on Soil Organic Matter in a Thin Black Chernozem in Western Canada. *Soil Biol Biochem* 1991a; 23:443.
35. Beri V, Sidhu BS, Bahl GS, *et al*. Soil Use Manage 1995; 11:51–54.
36. Manna MC, Swarup A, Wanjari RH, *et al*. *Field Crops Res* 2006; 93:264–280.
37. Rudrappa L, Purakayastha TJ, Singh D, *et al*. *Soil Till Res* 2006; 88:180–192.
38. Subehia SK, Verma S, Sharma SP. *J Indian Soc Soil Sci* 2005; 53:308–314.
39. Sharma MP, Bali SV, Gupta DK. *J Indian Soc Soil Sci* 2006; 48:506–509.
40. Balaguravaiah D, Adinarayana G, Prathap S. J Indian Soc Soil Sci 2005; 53:608–611.
41. Swarup A, Manna MC, Singh GB. In: Adv Soil Sci, CRC Lewis Publishers, Boca Raton, FL 2000; p261–281.
42. Gupta RK, Singh Y, Singh V. *Indian Farming* 2006; 56(4):10–14.
43. Tolanur SI, Badanus VP. *J Indian Soc Soil Sci* 2003; 51:41–44.
44. Yaduvanshi NPS. *J Indian Soc Soil Sci* 2001; 49:714–719.
45. Vermani SM, *Bull Indian Soc Soil Sci* 1994; 16:101–106.
46. Das SK, Hnal C. *J Indian Soc Soil Sci* 1994; 16:92–100.

47. Purakyastha TJ, Chhonkar PK, Bhadraray S, *et al*. *Aust J Soil Res* 2007; 45:33–40.

48. Dhaliwal SS. PhD. Thesis, 2003; PAU. Ludhiana.

49. Ernstings A, Rughani D. Climate Geo-engineering with "Carbon Negative" Bioenergy 2008; Biofuelwatch. http ://www.biofuel watch.org.uk./docs/cnbe/cnbe.html.

50. Lehmann. Stability of black carbon/biochar presentation at SSSA conference, October, 2008; htt://www.biochar-international. Org/sssa2008presentation.html.

51. Nguyen. Long-term black carbon dynamics in cultivated soils. Biogeochemistry 2003; 89:295–308.

52. Kumar P, Verma TS, Sharma PK. *J Indian Soc Soil Sci* 2006; 54:485–488.

53. Chand S. Integrated Nutrient Management for Sustaining Crop Productivity and Soil Health. International Book Distributing Company, Lucknow, India 2008; 112.

54. Chand S. Integrated Nutrient Management in Mustard. PhD Thesis, MPUAT, Udaipur, India 2001.

55. Chand S, and Pabbi S. Vermi-composting in organic farming in souvenir of agriculture summit-2006. Organized by Ministry of Agriculture GOI and FICCI, Vigyan Bhavan, New Delhi 2006; p1–6.

Carbon sequestration and farming are the best tool to tackle global warming and for cooling the Earth. — Subhash Chand

CHAPTER 5 Resource Conservation Technologies (RCTs) for Sustainable Agriculture and Resource Conservation

Conservation tillage cannot be adopted in isolation; it is a basic management tool. — Lal 1989

The indiscriminate use, rather misuse of natural resources especially water has led to the groundwater pollution as well as depletion of ground water resources. Presently, we are sitting on the volcano and if the situation is not improved we will face water wars in the near future, the sign of which are quite visible in the surface water dispute between Punjab and Haryana and some other states in India. Depleting soil organic carbon status, decreasing soil fertility and reduced factor productivity are other issues of concern. These evidences indicate that rice-wheat system has weakened the natural resource base. If we continue to exploit the natural resources, the productivity and sustainability is bound to suffer. Therefore, in order to meet the aim of sustainable yields over time it is the need of the hour to avoid further degradation of the natural resources. Moreover, in the face of World Trade Organisation (WTO) regime, we must produce at lower cost to be competitive in the international market being India already a surplus state in food-grain production. To meet these needs, the agricultural system must develop cost effective technologies suitable for harnessing the untapped potential especially in the North Eastern parts of the Indo-gigantic plains (IGP). The resource conservation technologies to economise on cost of production and need for conservation agriculture has assumed lot of significance. During the World Food Summit of 2002, the world leaders vowed to reduce this number to half by 2015. The UN Millennium Development Goals called for ensuring food as well as environmental security for all and can be achieved by increasing the crop productivity per unit of inputs (seed, fertilizer, water and land, etc.) and there is an urgent need to arrange a square meal for every one without deteriorating natural resource base.

5.1 CONSERVATION AGRICULTURE (CA) RELEVANCE

Conservation tillage methods are much more than just reducing the mechanical tillage. In a soil that is not tilled for many years, the crop residues remain on the soil surface and produce a layer of mulch. This layer protects the soil from the physical impact of rain and wind, but it also stabilizes the soil moisture and temperature in the surface layers. Thus, this zone becomes a habitat for a number of organisms, from larger insects down to soil borne fungi and bacteria. Those organisms macerate the mulch, incorporate and mix it with the soil and decompose it so that it becomes humus and contributes to the physical stabilization of the soil structure. At the same time this soil organic matter provides a buffer function for water and nutrients. Larger components of the soil fauna, such as earthworms, provide a soil structuring effect producing very stable soil aggregates as well as uninterrupted macropores leading from the soil surface straight to the subsoil and allowing fast water infiltration in case of heavy rain events. This process carried out by the edaphon, the living component of a soil, can be called "biological tillage". However, biological tillage is not compatible with mechanical tillage and with increased mechanical tillage the biological soil structuring processes will disappear. Certain operations such as mould board or disc ploughing have a stronger impact on soil life than others as for example chisel ploughs. Most tillage operations are, however, targeted at a loosening of the soil which inevitably increases the oxygen content in the soil leading to mineralization and thus to a reduction of the soil organic matter which is, at the same time substrate for soil life. Thus agriculture with reduced mechanical tillage is only possible when soil organisms are taking over the task of tilling the soil. This, however, leads to other implications regarding the use of chemical farm inputs. Synthetic pesticides and mineral fertilizer have to be used in a way that does not harm soil life.

As the main objective of agriculture is the production of crops, changes in the pest and weed management becomes necessary. Burning of plant residues and ploughing of the soil is mainly considered necessary for phytosanitary reasons controlling pests, diseases and weeds. In a system with reduced mechanical tillage based on mulch cover and biological tillage, alternatives have to be developed to control pests and weeds. Therefore, "Integrated Pest Management" becomes mandatory. One important element to achieve this is crop rotation, interrupting the infection chain between subsequent crops and making full use of the physical and chemical interactions between different plant species. Synthetic chemical pesticides, particularly herbicides, are in the first years inevitable, but have to be used with very much care to reduce the negative impacts on soil life. To the extent that a new balance between the organisms of the farm-ecosystem, pests and beneficial organisms, crops and weeds, becomes established and the farmer learns to manage the cropping system, the use of synthetic pesticides and mineral fertilizer tends to decline to a level below the original "conventional" farming.

Therefore, although the entry point is a reduction of mechanical soil tillage, "Conservation agriculture" (CA) involves a complete change in the crop production system. It involves modifications in the machinery, which means more mechanisation, maintenance of surface residues providing at least 30% soil cover, minimum soil disturbance, adjustment, if required, in the cropping system, minimum and need based use of chemicals[1].

5.1.1 Goal of CA/Long-term Objectives

Conservation agriculture aims to conserve, improve and make more efficient use of natural resources like soil, water and biological resources combined with inputs. It contributes to environmental conservation as well as to enhanced and sustained agricultural production. It can also be referred to as resource-efficient/resource effective agriculture.

5.1.2 Peculiarities of CA

Conservation agriculture maintains a permanent or semi-permanent organic soil cover. This can be a growing crop or dead mulch. Its function is to protect the soil physically from sun, rain and wind and to feed soil biota. The soil microorganisms and soil fauna take over the tillage function and soil nutrient balancing. Mechanical tillage disturbs this process. Therefore, zero or minimum tillage and direct seeding are important elements of CA. A varied crop rotation is also important to avoid disease and pest problems.

Rather than incorporating biomass such as green manure crops, cover crops or crop residues, in CA this is left on the soil surface. The dead biomass serves as physical protection of the soil surface and as substrate for the soil fauna. In this way mineralization is reduced and suitable soil levels of organic matter are built up and maintained.

5.1.3 Everything is not CA

Zero tillage: Zero tillage is a technical component used in Conservation agriculture, but not everyone carrying out zero tillage is using Conservation Agriculture. Conservation agriculture not only avoids tillage by forcing the seed with appropriate direct drills into the soil, by maintaining a soil cover it also improves the structure of the soil. This facilitates direct planting. Conservation agriculture uses biological tillage. Zero tillage as stand alone technique can also be applied in conventional agriculture under certain circumstances.

5.1.4 Conservation Tillage Practices (CTPs)

Conservation tillage is a practice that leaves crop residues on the surface, which increases water infiltration and reduces erosion. It is a practice used in conventional agriculture to reduce the effects of tillage on soil erosion. However, it still depends on tillage as the structure-forming element in the soil. Nevertheless, conservation tillage practices such as zero tillage practices, can be transition steps towards conservation agriculture.

5.1.4.1 Direct planting/seeding/sowing

This is only a technique that refers to seeding/planting without preparing a proper seedbed. The same equipment is used in conservation agriculture. However, the term direct seeding can also be used for implements, which combine primary and secondary tillage and seeding in one machine/tractor operation like the rotary till drills.

5.1.4.2 Organic farming in relation to CA

Conservation agriculture is not a synonym of organic farming, although it is based on natural processes. CA does not prohibit the use of farm chemical inputs. For example, herbicides are important components in conservation agriculture, particularly in the transition phase, until the new balance of weed populations is managed. However, in view of the importance of the soil life for the system, farm chemicals, including fertilizer, are carefully applied and over the years, quantities applied tend to decline. In some cases organic farming can be practised within the CA framework[1].

5.1.4.3 Is CA compatible with IPM?

Conservation agriculture is not only compatible, but actually works on IPM principle. CA, like IPM, enhances biological processes. It expands the IPM practices from crop and pest management to land husbandry. Without the use of IPM practices the build up of soil biota for the biological tillage would not be possible.

5.1.4.4 What is the role of animal husbandry in CA?

Livestock production can be fully integrated into conservation agriculture, by making use of the recycling of nutrients. This reduces the environmental problems caused by concentrated intensive livestock production. Integration of livestock into agricultural production enables the farmer to introduce forage crops into the crop rotation, thus widening it and reducing pest problems. Forage crops can often be used as dual-purpose crops for fodder and soil cover. Particularly in arid areas with low production of biomass, the conflicts between the use of organic matter to feed the animals or to cover the soil has still to be resolved.

5.1.4.5 What are the downsides of CA?

CA may require the application of herbicides in the case of heavy weed infestation. During the transition phase, certain soil borne pests or pathogens might create new problems due to the change in the biological equilibrium. Once the conservation agriculture environment has stabilized it tends to be more stable than conventional agriculture. So far, there has been no pest problem that could not be overcome in conservation agriculture.

5.1.4.6 Benefits of CA

Conservation agriculture attracts different people for different reasons:

Farmers:

- Reduction in labor, time, farm power
- Reduction in cost

- In case of mechanized farmers: longer lifetime and less repair of tractors, less power and fewer passes, hence much lower fuel consumption
- More stable yields, particularly in dry years
- Better traffic ability in the field
- Gradually increasing yields with decreasing inputs
- Increased profit, in some cases from the beginning, in all cases after a few years

Communities/environment watershed:

- More constant water flows in the rivers, re-emergence of dried wells. Cleaner water due to less erosion
- Less flooding
- Less impact of extreme climatic situations (hurricanes, drought, etc.). Less cost for road and waterway maintenance
- Better food security

At global level (world view)

5.1.4.7 Carbon sequestration (Greenhouse effect)

In some places no-till farmers start to receive carbon-grant payments; the global potential of conservation agriculture in carbon sequestration could equal the human made increase in CO_2 in the atmosphere.

- Less leaching of soil nutrients or chemicals into the ground water
- Less pollution of the water
- Practically no erosion (erosion is less than soil build up)
- Recharge of the aquifers through better infiltration
- Less fuel use in agriculture

5.2 ISSUES IN CA

Despite its advantages, CA has so far spread relatively slowly for a number of reasons. Firstly, there is greater pressure to adopt conservation agriculture in tropical, rather than temperate climates. It has taken a long time, but over the past 20 years the establishment of a local knowledge base has ensured its spread. In some states of Brazil it is official policy, in Costa Rica the Ministry of Agriculture has a Department for Conservation Agriculture, so in these cases the policy makers have been convinced. The adoption of CA in the US was probably due to a mixture of public pressure to fight erosion and the financial incentives of reduced tillage. Europe is slowly getting there — farmers still do not feel sufficient pressure and environmental indicators (erosion, flooding) are not yet taken seriously enough. In India, farmers are moving towards zero tillage/conservation agriculture mainly due to increased costs of production especially that of fuel for tractors and other agricultural inputs and to some extent to increasing problem of water availability due to falling ground water table[1]. Ground water level is depleting due to overexploitation of ground water and further problems aggravated by unpredictable weather and rainfall pattern.

CA has great potential in Africa due to its propensity to control erosion, gives more stable yields and reduces labor. There are a number of ongoing initiatives promoting different practices, from conservation tillage up to conservation agriculture. Another vast area where the adoption of CA would be extremely beneficial is Central Asia. In the countries of the former USSR conventional agriculture is virtually impossible because of environmental problems (erosion) and because of a lack of farm machinery, which has to be replaced. Unless conservation agriculture is adopted, the investment in new machinery will have to be very high. India can also be immensely benefited by adopting conservation agriculture by reducing the cost of production and be more competitive in the global market.

Converting to conservation agriculture needs higher management skills, the first years might be very difficult for the farmer; therefore she/he might need moral support (from other farmers or from extension services) and perhaps even financial support (to invest into new machinery like zero tillage planters). As it requires a complete change of understanding, the scientific and technical sectors often do not support conservation agriculture, fearing that they would contradict themselves.

5.2.1 Necessary Technologies are Often Unavailable

In order to try CA, the minimum a farmer needs is a zero tillage planter, which might not be available in the neighbourhood. Buying one without knowing the system or even having seen it, is a risk that few farmers take. Machinery dealers might not wish to promote CA as long as it is not supported by extension. This is partly due to the cost of the equipment, but more importantly because the widespread adoption of CA will reduce machinery sales, particularly of large tractors.

5.2.2 Is Conservation Agriculture Real?

Conservation agriculture is being used on more than 45 M ha, mostly in South and North America. Its use is growing exponentially on small and large farms in South America, due to economic and environmental pressures. Farmers using CA in South America are highly organized (in regional, national and local farmers organizations), and are supported by institutions from North and South America. In Europe the European Conservation Agricultural Federation, a regional lobby group, has been founded. This body unites national CA associations in the UK, France, Germany, Italy, Portugal and Spain. In India also, during the last ten years zero tillage has increased from nil to more than 2 M ha and is spreading very fast. Now the emphasis is being shifted from conservation/zero tillage to conservation agriculture and a few machines are in the testing stage that can seed into loose residues left after combine harvesting of the crop. (Subhash Chand *et al.*, 2012)

5.3 NEW MACHINES FOR CA

Efforts are being made to develop and fine tune suitable machines for seeding into loose residues left after combine harvesting. The need is felt due to depletion of soil organic carbon and causing environmental pollution due to burning of crop residue after combine harvesting. At present four machines are under testing and evaluation for seeding direct seeded rice and wheat, which are briefly discussed hereunder[2].

5.3.1 Double Disc Coulters

This is one of the second-generation machines being tried under loose residue conditions. It has double disc coulters fitted in place of tynes to place the seed and fertiliser into the loose residues. The problem being faced with this machine is that being light weight it fails to cut through the loose residues and the seed and fertiliser is dropped on the top of it, part of which reaches the soil surface. Irrigation is required immediately after seeding in order to facilitate the germination. This machine may work up to a residue load of about 4 to 5 Uha.

5.3.2 Punch Planter/Star Wheel

This is another machine, which is being tested for seeding into the loose residues. This mechanism is being used widely around the world under non-rice situations, but its utility under rice wheat system with a residue load of 6 to 10 Uha is still to be proved. The initial results indicate that it may work under low residue load of up to 3 Uha. At present this machine drops the fertiliser on the surface in front of the moving star wheels, which is not the proper method of placing the fertiliser.

5.3.3 Happy Seeders

This is another machine, which cut and lift the residue in front, place seed and fertiliser using zero till machine and drop the chopped straw behind on to the seeded area. This machine is capable of seeding into the loose residue load of up to 10 Uha. Recently an improved version of this machine has been developed which cuts in strips of 5 cm only in front of tynes, which reduces the energy requirement of the machine.

5.3.4 Rotary Disk Drill (RDD)

This machine is based on the rotary till mechanism. The rotar is a horizontal transverse shaft having six to nine flanges fitted with straight discs for cutting effect similar to the wooden saw while rotating at 220 RPM. The rotary disc drill is mounted on the three point linkage system and is powered through the power take-off (PTO) shaft of tractor. The rotating discs cut the residue and simultaneously make a narrow slit into the soil to facilitate placement of seed and fertilizer. The machine can be used for seeding under conditions of loose residues as well as anchored and residue free conditions. If the machine is to be used under loose residue condition, it is better to use an offset double disc assembly for placement of seed and fertilizer otherwise inverted T-type or chisel type openers can also be used. The rotary disk drill can also be easily converted into rotary till drill by replacing the discs with L or J-shaped blades on the rotor. The rotor completely pulverizes the soil leading to a clean and fine tilth. In case rotary disk drill is to be used as a zero till drill, straight blades or discs can be used for minimal soil disturbance. However, it must be remembered that in presence of loose residues, combination of rotary disk with coulter double disk completely avoids the raking problem of residues during seeding operations. Thus the newly designed rotary disk drill is a multipurpose machine, which can be used to seed under diverse situations depending upon the presence and condition of crop residues.

The rotary disk drill can be used in manually as well as combine harvested fields for direct drilling of seed and fertiliser in a single tractor operation under variable field soil moisture conditions. Extensive field trials of the newly designed rotary disk drill are already underway to evaluate the performance and durability of rotary disks. The direct seeded rice and wheat crops were successfully established in loose residues up to 8 tonnes per hectare using rotary disc drill. The machine was also tested at the farmers' field for its capability to sow under zero till as well as under loose residue conditions and the result was very encouraging. However, one problem being faced was frequent blunting of the powered discs.

5.4 RCTs IN RECLAIMED ALKALI SOILS

Declining groundwater levels and deteriorating quality of soil in the Indo-Gangetic plains is emerging a serious concern for agricultural sustainability in the near future. Depletion of groundwater in areas irrigated by tube wells and associated water quality concerns have brought about heightened awareness of the need for the judicious use of rain, surface and ground water resources[2]. The pressures on natural resources are immense. Soils are depleting in their fertility as a result of continuous and intensive cropping. The organic carbon levels in Punjab soils have decreased from 0.5% (1950–60) to about 0.25% at present. Tillage costs are rising, which accentuates the already serious labor shortages during peak periods of land preparation and harvest. For these and other reasons, the long-term sustainability of these systems is now a subject of attention. Further, increased cost of cultivation and declining productivity is compelling the farmers to quit the farming[2].

There is a general consensus that quality of natural resources base needs to be improved for enhanced productivity, sustainability and profit. Also, it is believed that future productivity growth would come through efficient management of inputs (water, nutrients and energy) and better risk management strategies. Targeted resource conserving technologies offer newer opportunities for better livelihood for the resource poor, small and marginal farmers of the region.

Though rice-wheat systems are critical in Indo-Gangetic plains, yet valuable information has remained underutilized by the farmers. The improved tillage, residue management and crop establishment practices show real potential for improving the productivity and profitability of rice-wheat systems. Reduced and zero tillage can improve yields, increase input-use-efficiency, reduced the intensity of machinery use and lower the production costs. Therefore, efficient soil and water management practices, such as tillage, irrigation and nutrients have to be fine-tuned according to the crop establishment requirements in a particular transect. Keeping the above facts in view, a long-term experiment is started at CSSRI to address some of the problems faced by the farmers in managing the natural resources for increasing the rice-wheat productivity.

5.4.1 Immediate Objectives

- To study the effect of resource conservation options on water, nutrient, energy-use-efficiency and crop productivity
- To monitor the changes in physical, chemical and biological properties of soil as influenced by different resource conservation options
- To work out salt, water, nutrient, energy and gas fluxes in selected treatments to monitor impacts of conservation agricultural practices in moderating climatic variations
- To monitor changes in plant growth, yield and economics in different treatments

5.4.2 Long-term Objectives/Goals

- To evaluate different resource conservation options for the sustainability of rice-wheat cropping system
- To recommend the most profitable, eco-friendly and resource upgrading technology to the farmers of the region
- To provide input for making suitable policy for the management of natural resources for rice-wheat cultivation in Indo-Gangetic plains

5.5 RESOURCE CONSERVING TECHNOLOGIES — POTENTIAL TOOLS FOR ATTAINING FOOD, NUTRITIONAL AND LIVELIHOOD SECURITY

In early seventies, an estimated 920 million people in the developing world were chronically undernourished, with insufficient food but at the beginning of nineties, despite continuing population growth, the figure had been reduced to 840 million about 20% of the total population of the developing world. Out of these, a major part lives in developing and poor countries in Asia and Africa Continents. But this will require a huge investment (public as well as private) in agriculture and allied sectors which are critical factor in achieving the goal. In view of this, top most priority should be given for investment in agriculture[4].

The International Food Policy Research Institute (IFPRI) estimated that world food production will grow by an average of 1.5% per year between 1990 and 2020, if investments such as in agricultural research, infrastructure, irrigation, markets and extension and training are maintained at least at 1980s levels. In India, agricultural and allied sectors contribute 25% of GDP which affects the overall GDP by 0.5%–1.0%. As per our National Agriculture Policy, we should achieve 4.0% growth rate but it is hovering around 1.50%–1.75%, which cannot ensure the food security of the people[2].

Tremendous changes have occurred in the world economy over the past two decades, but external development assistance has declined and agriculture has been disproportionately affected. Total external commitments to agriculture in 1994 were 23% below those of 1980. Food aid has also declined from almost 17 million tonnes (cereal equivalent) in 1992–93 to around 9 million tonnes in 1994–95. For example, the investment in irrigation has slowed, especially in Asia. In India, since the mid 1990s private investment in agriculture has stagnated while public investment has continued to decline. It is also essential to increase long-term public investment in agriculture, roads, telecommunications, electricity and irrigation, which will stimulate private investment and contribute to a revival of growth momentum in agriculture.

Massive donors backing are required for enhancing the investment in agricultural development, research, education and extension activities. Presently World Bank, International Monetary Fund (IMF), FAO, International Fund for Agricultural Development (IFAD), Asian Development Bank (ADB), USAID, and Department for International Development (DFID) are generally providing support to different projects in raising the status of food security of the people for developing and least developed countries. In such a situation, biotechnology, post-harvest technology (PHT) research and peri-urban agriculture should be given priority to increase production to feed the growing population within the shortest possible time.

Crop diversification, livestock raising, dairy production, and fisheries cultivation, these are all enthusiastic effect to build a reservoir of food for future food security of the population of the country. A carefully plan peri-urban agriculture had the potential to put agriculture on the path of selfpropelled and self-reliant development under leadership through an indigenous public private partnership in the areas of food security and poverty[4].

Farm mechanization through resource conserving technologies (RCTs), viz. zero tillage, bed planting, mechanical rice transplanting and laser leveling are some of the efficient tools for increasing the crops production and their productivity and reducing the cost of production.

In South Asia alone, there is about 14.0 million ha land which remains fallow after rice crop (rice fallows), mostly due to late harvesting of rice crop, poor irrigation facilities, excessive wetness and poor drainage. If these lands can be brought under cultivation, by adopting these technologies, to some extent can solve the food, nutritional and livelihood security problem of Asian countries and also generate additional employment opportunities for millions of inhabitants of this region. A major part of this falls in India alone (81.5%), out of which about 50% falls in Eastern IGP region.

Crop establishment methods are exceedingly important factor in improving the productivity of crops. Land preparation cost represents a major portion of cost and with ever increasing oil prices. There is an urgent need to reduce the tillage practices for reducing the same and also the cost of fertilizers and water[4].

5.6 RCTs IN RICE-WHEAT SYSTEM AND NEED FOR CONSERVATION AGRICULTURE

Earlier, agriculture was focused on achieving food security through increased coverage under high yielding varieties, expansion of irrigation and increased use of external inputs. This enabled the rice-wheat (RW) to emerge as a major cropping system in the Indo-Gangetic plains (IGP) leading to the Green Revolution. These two crops together contribute more than 70% to the total cereal production in India. The estimated area of RW system in India is around 10.0 M ha. At present, the food situation is comfortable but increasing production to meet the needs of ever growing population is full of uncertainties. With the increase in population, more and more land will be required for urbanization and productivity needs to be increased to meet the domestic and industrial demand[4].

The resource conservation technologies to economise on cost of production and need for conservation agriculture are briefly discussed hereunder.

5.7 LASER LAND LEVELLING (LLL)

Laser land levelling is the process of smoothening the land surface within 2 cm from the average elevation of the field using laser equipped bucket which scraps from higher places and spread onto the low lying areas. This technology is a prerequisite for enhancing the benefits of other resource conservation technologies.

Generally, fields are not properly levelled leading to poor performance of the crop, because, part of area suffers due to water stress and part due to excess of water. After laser levelling the field, it has been observed that yield enhances from 10 to 25 per cent. The higher yields are due to uniform crop stand, water distribution, crop growth and maturity. In addition to higher yield, this technology saves 35–45 per cent water, a scarce resource, due to higher application efficiency, increases nutrient use efficiency by 15–25 per cent, reduces weed problem and increases the cultivable area by 3 to 6 per cent due to reduction in area required for bunds and channels.

5.8 RESOURCE CONSERVATION TECHNOLOGIES (RCTs) IN WHEAT

5.8.1 Zero Tillage

This is a resource conservation technology in which wheat is directly seeded into the undisturbed soil after rice harvesting using a specially designed machine. In this seed and fertiliser is placed into narrow slits created by the knife type furrow openers of zero tillage ferti-seed drill. This technique was first adopted in the high yielding, more mechanized areas of North Western India and Pakistan where lot of money was being invested on field preparation. This technology provided an opportunity to reduce the cost of cultivation by ₹ 2500–3000 per hectare thereby increasing the profit margin of the farmers. In addition, development of resistance against commonly used herbicide 'isoproturon' in *Phalaris minor* was also responsible for its adoption in rice-wheat system due to lower incidence of this weed under zero tillage[3].

The work on zero tillage was initiated with the development of zero tillage machines by Dr Bachan Singh and his group during 1992–93. The first version received at DWR had problem of patchy germination due to rigid side drive wheel, which slipped, at places near the bunds as well as where depressions were made due to tractor or combine harvester tracks and even human foot prints. Replacing the side drive wheel with balancing wheel and bringing the drive wheel to front and making it floating type solved this problem. The area under zero tillage was negligible till 1996–97 which increased slowly to 0.2 M ha during 2001–02. Thereafter the increase was very fast with less than 0.5 M ha in 2002–03 to around 2.0 M ha during 2004–05, including the area under reduced tillage. This speed of adoption was possible only because of farmer's participatory approach adopted by the scientists working on this technology.

5.8.2 Energy and Economics

Considering the total cost of field preparation and drilling or broadcast sowing of wheat, it was found that the maximum cost was ₹ 1637 for broadcast sown wheat followed by drill sown (₹ 1413) and minimum was for zero tillage which was only ₹ 179/ha. The total energy required for various tillage options varied from 20279 MJ/ha for zero tillage to 23631 MJ/ha for broadcast sown wheat. The benefit cost ratio was highest for zero and lowest for broadcast sown wheat, whereas, the specific energy (energy spent per kg of biomass production) requirement was lowest for zero and highest for broadcast sown wheat after conventional field preparation.

5.8.3 Reduced/Minimum Tillage

The impact of zero tillage is that most of the farmers have shifted from intensive tillage undertaking 6 to 12 tractor operations to reduce tillage involving 2 to 3 operations with various farm implements. Reduced tillage has advantage over conventional tillage as it saves on tillage cost with similar crop productivity. Reduced tillage has no apparent advantage over ZT but farmers are sometimes forced to undertake 2 to 3 tractor operations due to the following reasons:

- With some of the farmers it is a problem of mind set as they think that fields do not look tidy in the initial stages
- Early transplanting or using early maturing varieties of rice vacates the fields by 15th week of October whereas sowing of wheat is generally done from 15th week of November. In such fields standing stubble of rice become loose which creates problem in smooth running of ZT machine

- Some broad leaved weeds like *Rumex* spp. germinate in the month of October after harvest of rice. Under such situations spray of glyphosate is recommended in zero tillage sowing of wheat. But it costs higher than 1–2 ploughing and also due to lack of proper spray technique some weeds are left after spray. Therefore, farmers prefer 1–2 tillage instead of spraying
- The field becomes uneven due to formation of tracks in wet soil after combine harvesting. This happens either due to late irrigation or if rain occurs toward the end of rice season as it happened during 2004. Under such situation, it becomes necessary to level those patches with 1 or 2 harrowing

5.8.4 Rotary Tillage

The machine is a combination of rotavator, seed-cum fertiliser drill and a light planker-cum-driving wheel at the back. The machine uses a nine-row standard seed-cum-fertiliser drill, in addition to a rotary tiller. As the name of the machine suggests, it tills the soil, rather completely pulverises the top 8 to 10 cm, before placing seeds and fertilisers. The rotary tiller is a horizontal transverse shaft fitted with L-shaped blades. It is powered through the power take-off (PTO) shaft fitted at the rear of every tractor. The rotor completely pulverise the soil along with the cutting and mixing of residue or weeds present in the field leading to a clean very fine tilth. The machine has 7 gangs of six blades each and can be operated by a tractor of about 40 hp or more in dry/optimum moisture condition or about 30 hp or more for puddling of rice fields. The rotary-till drill can be used in manually harvested rice field at optimum soil moisture conditions for directly seeding wheat and fertiliser placement in a single tractor operation. This saves more than 70 per cent of the time and energy compared to conventional field preparation being followed by the farmers with 7 to 10 per cent higher yield. In addition, it helps in advancing the sowing in case the wheat seeding is getting delayed beyond 25th November. In combine harvested rice fields the machine can be used after removal or burning of the loose paddy straw. The field preparation by rotary tillage is better than conventional as the soil is completely pulverised. Incorporation of green manure crops, weeds and crop residues can also be accomplished in a single operation. As the seeding is done simultaneously, the soil moisture is conserved leading to better crop stand in addition to savings on energy, time and labor requirement making this technology more economic and eco-friendly.

The machine was initially aimed to sow wheat directly after paddy harvesting. Subsequently, it was also tried for puddling of rice field since the rotary tillers are known to be excellent puddlers. Seed hoppers and seed placing units can be detached from the frame fixed on the rotary unit and the rotary unit alone can be used for puddling. A single operation after ponding of water was found sufficient for puddling followed by rice transplanting after 1 to 3 days depending upon soil type.

5.8.5 Bed Planting

This is a water saving technology, which saves 30 to 40 per cent water for growing wheat depending upon the soil type. In addition to water saving this technology has numerous advantages in rice-wheat system. Although there is no saving on the cost of land preparation or time but it can become cost effective by using the same beds for rice without reshaping. In this technology after preparation of land all three activities named bed formation, placement of fertilizer and sowing of wheat are done in single operation. Crop cultivars are known to vary significantly in their performance on raised beds. It is suitable for seed production also because of bolder grain and easier rouging. It reduces the herbicide dependence due to mechanical weed control with the same bed planter fitted with inter culture tines with simultaneous placement of fertiliser. In situations where sowing can be delayed due to pre-sowing irrigation, dry seeding can be done on raised beds followed by irrigation immediately after seeding. Irrigation can also be given at grain filling stage, which is generally avoided by the farmers for fear of crop lodging. In this technology, nitrogen, use efficiency is also higher because of light irrigation and top dressing on beds.

5.8.6 Surface Seeding

This is a technology, which does not require any field preparation and sowing of wheat is done in standing rice crop a few days before or immediately after rice harvesting. There are areas in the Eastern part of Indo-Gangetic plains where land remains wet after rice harvesting for a long time and field preparation for sowing second crop is not possible. Under such conditions surface seeding provides an opportunity to take wheat crop in *rabi* season. Even in areas where field preparation is possible, wheat sowing is delayed leading to very low yields. So by adopting surface seeding one can harvest higher yields. In this technology dry or soaked seed is broad casted over the wet soil. To prevent bird damage the seed is invariably coated with cow dung. For proper and uniform crop stand drum seeders can also be usefully employed after rice harvesting. Nowadays, farmers are practising the surface seeding successfully not only in wheat, but also in other upland crops like pea, gram, lentil, etc.

5.8.7 Effect of Tillage Options on *Phalaris*

The lower *Phalaris minor* population and dry weight was observed under zero tillage and the higher under conventional tillage system. The less weed problem under zero tillage may be due to less soil disturbance helping in keeping the weed seeds at depth from where it could not germinate. Unchecked weed growth during the crop season caused maximum yield loss in conventional tillage, followed by furrow irrigated raised bed-planting system (FIRBS) and the minimum reduction was observed in zero tillage. Therefore, ZT seems to be a cost effective and a sustainable weed management option.

5.8.8 Water Use and Savings under Various Tillage Options

There are two schools of thoughts among the scientists, one who says that zero tillage can save up to 25 per cent water whereas the other group feels that such a huge saving is not possible although there may be some savings on account of comparatively lower infiltration rate and subsequently higher retention of moisture in zero till sown fields. The argument the first group put forward is that wheat can be sown without pre-sowing irrigation. This argument doesn't carry much weight because the soil can be prepared using the same moisture and wheat can be sown almost simultaneously even under conventional field preparation. If the scenario of rice residue incorporation followed by heavy irrigation of about 10 cm to decompose it is considered, then there is a possibility of water saving up to 20–25 per cent. To address this issue a study was conducted during rabi 2001–02 season involving wheat sowing under three tillage options (ZT, FIRBS and CT) in manually harvested (near to ground) rice field. A pre-germination irrigation was applied after 3 days of seeding in FIRB system for ensuring germination and good crop stand. Thereafter, need-based irrigations were applied. The water applied at each irrigation was measured using a Parshal flume. At each irrigation, water applied was marginally lower under zero tillage whereas in FIRBS it was only around 35 to 40 per cent compared to conventional tillage. The total post seeding irrigation water requirement was about 25 cm in conventional tillage whereas zero tillage required about 3 per cent less irrigation water. The irrigation water saving was more than 30 per cent in case of FIRB system of wheat cultivation.

The average yield of wheat in various tillage options varied from 60.35 q ha^{-1} in FIRB system to 64.06 q ha^{-1} under zero tillage. The total water use calculated by accounting for the rainfall and profile depletion varied from around 30 cm in FIRBS to about 37 cm in conventional field preparation whereas zero tillage required about 36 cm of water. The total water saving compared to conventional was 2.58 per cent in zero tillage and about 20 per cent in FIRB system of wheat cultivation. The water use efficiency was highest in FIRBS and the lowest in conventional system of field preparation. It can therefore be summed up that there is no great water saving or usage efficiency under ZT even though numerically there is a minor advantage.

5.8.9 Eco-friendly Tillage Options (EFTOs)

These alternate tillage technologies are also environmental friendly compared to traditional broadcast sown wheat. The carbon dioxide emission due to burning of fuel (assuming 2.6 kg CO_2 production/litre of diesel burnt) during field preparation was 208.00 kg/ha in conventional farmers' practice and was only 15.60 kg/ha in zero tillage and 36.92 kg/ha in rotary tillage. Even conventional field preparation followed by drill or bed planter sowing on FIRBS resulted in reduction of carbon dioxide emission by 18%–19% compared to broadcast sown wheat. This corresponding reduction in CO_2 emission comes to 82.25% and 92.50% for rotary and zero tillage, respectively.

5.9 RCTs IN RICE FOR WATER SAVING

Wet tillage, i.e. puddling for growing rice has been suspected to adversely affect the soil and water resources and efforts are being focused on developing and fine tuning the technologies to grow direct seeded rice or transplanted rice without wet tillage. The initial results are encouraging and are discussed briefly here. Direct dry seeded and unpuddled transplanted rice: In this system rice is grown like any other upland crop with seed placed in the soil by seed cum fertilizer drill with or without ploughing. The traditional practice of growing rice, i.e. transplanting in puddle conditions requires higher amount of water and labor. In puddle soil once the water dries, cracks develop and water percolates beyond the root zone along with nutrients. Generally, plant population is 18–20 in manual random transplanting against the recommended density of 35–40 plants, which is a major constraint in achieving higher yield of transplanted rice[3].

Direct seeding has advantages of faster and easier planting, reduced labor and less drudgery with earlier crop maturity by 7–10 days, more efficient water use and higher tolerance of water deficit, less methane emission and often higher profit in areas with an assured water supply. Thus the area under direct seeded rice has been increasing as farmers in Asia seek higher productivity and profitability to offset increasing costs and scarcity of farm labor. Weed control is a major issue in direct seeded rice and to overcome these problem intensive efforts are being made by the agricultural scientists. In some soils, spray of micronutrient like Zn and iron may be needed to remove their deficiency. Direct seeding of rice using zero till drill, rotary till drill, drum seeder as well as broadcasting under various field preparation or puddling options was tried at DWR research farm. Seeding depth was kept at 2–3 cm while using drill for seeding. For comparison purposes transplanting was also done under conventional puddling as well as under zero tillage and after field preparation with rotary tiller. The rice variety used was IR 64. Direct seeding was done in the first week of June on the same day when nursery was sown for transplanting. For weed control Sofit @ 1500 ml/ha was applied after four days of direct seeding followed by one weeding at around 35 days after seeding.

Among the direct seeding options, the yield recorded was highest where rice was seeded using rotary till drill followed by broadcasting sprouted rice seed after field preparation by rotary tillage and lowest when broadcasted under zero tillage. The mean yield in rotary tillage was significantly higher compared to zero tillage. Direct drilling by zero till drill and rotary till drill was at par and as good as transplanting under zero tillage or after field preparation by rotary tillage and was significantly higher than drum seeding or broadcasting under zero tillage. Among transplanting and direct seeding options, highest yield was recorded in machine transplanting, which was significantly better than broadcasting and drum seeder, but statistically at par with other transplanting or seeding options. The yield was marginally higher in conventionally puddled conditions compared to transplanting without tillage, after field preparation by rotary tillage or direct drilling by zero or rotary till drill.

5.9.1 Direct Wet Seeded Rice

In this system sprouted seeds are broadcasted or placed with drum seeder under puddled or unpuddled conditions. Wet seeded rice also reduces labor costs and effective herbicides for weed control has helped making this technology more popular. Seed rate in drum seeded rice varies from 50–75 kg/ha whereas in broadcasting method of seeding 20–30 kg/ha is sufficient. In wet seeded rice puddling can be avoided without any adverse effect on rice yield. The observations at farmer's field showed that mortality of sprouted seeds are higher under puddled compared to unpuddled conditions.

A field trial on direct seeded rice was conducted with different seed rates varying from 30 to 80 kg/ha during 2002. Similar yield was recorded at varying seed rates suggesting that the seed rate can be further reduced. In 2003 rice season, an additional treatment of 20 kg/ha was included. The varying seed rates were kept based on earlier recommendation of the Directorate of Rice Research of 75–100 kg/ha. The variety used was IR 64 having a 1000 grain weight of about 26 grams. For a population of about 0.33 × 106 plants/ha recommended for transplanted rice, the seed requirement is likely to be around 11 kg/ha after giving an allowance of 20% loss in germination percentage of seed. If rodent and bird damage are further added to the estimates, almost double the seed requirement (20 kg/ha) should be good enough. The trial was sown in the first week of July during 2002 and second week during 2003 when the transplanting is generally done. The yield recorded was almost similar at seed rates of 20 to 80 kg/ha.

5.9.2 Weed Management in Direct Seeded Rice

Weed management is the major problem in direct seeded rice (DSR). Experiments have sown that it can be tackled successfully by integrated weed management practices which include stale bed technique, crop rotation, brown manuring, zero tillage, use of competitive varieties, water management, mulching, intercropping of cover crops and use of suitable chemicals at right time. Integrated weed management approach utilizes all suitable techniques and methods, which maintains the weed population below economic threshold level[3].

5.9.3 Leaf Color Chart (LCC)

Leaf color is a fairly good indicator of the nitrogen status of plant. Nitrogen use can be optimised by matching its supply to the crop demand as observed through change in the leaf chlorophyll content and leaf color. The leaf color chart developed by International Rice Research Institute, Philippines can help the farmers because the leaf color intensity relates to leaf nitrogen status in rice plant. The monitoring of leaf color using leaf color chart helps in the determination of right time of nitrogen application. Use of leaf color chart is simple, easy and cheap under all situations. The studies indicate that nitrogen can be saved from 10 to 15% using the leaf color chart.

5.10 CONCLUSIONS FOR ADOPTION OF RESOURCE CONSERVING TECHNOLOGIES

Techinologies (RCTs) leads to win win situations, reduce cost of production, improve production and productivity, save on energy, water and are leading to precision agriculture in an eco-friendly manner. For these reasons, these technologies are farmers best bet and have found favor with them and thus creating revolution in parts of the IGPs. Farmers should be reoriented, educated and trained properly for these technologies for their adoption.

There should be public awareness campaign of the benefits of RCTs at the farm, village, country and global level. There should be regular review, upgradation of mechanism of the technology on regular basis based on the feedback from the users for further improvement. There is a need to modify and develop more energy, water and nutrient efficient RCTs, made available to the farmers. By adopting RCTs, may lead to food, environment and livelihood security and lead to more employment generation.

For this, more investment (public and private) is needed in farm power and machinery sector of agriculture. More agro-industries should be established for manufacturing farm equipments at cheaper and affordable rates. For small farmers, the farm mechanization can be through cooperatives or panchayats on custom and hiring basis. This will lead to more and qualitative food production for ever-increasing hungry population led to more employment generation and ensures the livelihood on sustainable basis.

Corporate houses and multi-national companies (MNCs) might come forward for commercializing the agriculture through investments in contract farming for quality seed production, horticultural crops (fruits, vegetables and cut flowers), agro-industries (gatta factories, rice residues, pulp and paper), post-harvest for processing, and peri-urban agriculture.

National Commission on Farmers should be the apex body for advising the wiser investment in agriculture and allied sectors and similarly the recently constituted National Rainfed Area Authority should advise for proper investment in dry lands, which constitute about 2/3 or >60% but contributes very little to national food grain pool, for drought proofing though various means for improving the production and productivity of major crops like oilseed and pulses through crop intensification and diversification. This will lead to lesser imports of these commodities for which huge foreign exchange is drained and the saved money can be invested in agriculture and its allied sectors for enhancing the food, nutrition and livelihood security of the people.

5.11 KEY REFERENCES AND RESOURCES FOR FURTHER READING

1. CASA, Ground Water use in northwest India Culture for Advancement of Sustainable Agriculture. New Delhi, India 2004.
2. Sharma DP, Sharma SK, Joshi PK, *et al.* Resource Conservation Technologies in Reclaimed Alkali Soils. Technical Bulletin, Central Soil Salinity Research Institute Karnal, India 2008.
3. Chand S, Singh L and Singh P. Sustainable Agriculture, Food Security and Climate Change. Daya Publisher, New Delhi, 2012; p218.
4. Sharma RK. Sustainable Irrigated Agriculture Through Command Area Development. National level training course command area development with emphasis on land levelling shaping planning and design. Central Soil Salinity Research Institute, Karnal, India 2006; p153–167.
5. Tomar RK, Sahoo RN, Garg RN, *et al.* Resource Conserving Technologies Potential Tools for Attaining Food, Nutritional and Livelihood Security. *Indian Farming* 2006; p24–31.

RCTs should be popularised beyond rice-wheat cropping system for maximum production potential.— Subhash Chand

CHAPTER 6

Approaches in Fertilizer Recommendations (AFRs) for Maximizing Yield and Dynamic Soil Health

Man despite his artistic pretensions, his sophistication, and his much accomplishment over his existence to a 15 cm layer of topsoil and the fact that it rains. — Anon

The rational and balanced fertilisation leads to higher crop yields besides sustainable soil health. Soil testing refers to the chemical testing of soils for evaluating their fertility status with the object of making recommendations of fertilisers. It is prerequisite to know the nutrient imbalances in the soil and apply required amount of nutrients to correct imbalances and optimise crop nutrition. It also includes testing of soils for other properties like texture, pH, $CaCO_3$ content and parameters for ameliorating of chemically deteriorated soils for recommending soil amendments. Soil testing is the only tool known which helps to control soil fertility which is not a permanent or long lasting entity. In the known Indian history from 500 BC Indian soils supporting many kingdoms which flourished and perished resulting in improvised soils over centuries. Before Indian independence, farmers survived with extensive cultivation and low yields. After independence, substantial investments were made in creating irrigational potential and other infrastructural facilities throughout the country. Introduction of better varieties, improved cultural practices, improved drainage control of pests and diseases have helped to set the stage for higher yields. In addition to these huge investments were made on fertilisers. In future, production costs of N, P_2O_5 and K_2O fertilisers becoming costly and their demand is being depended upon the ability of the farmers to pay for them. Economic rationality therefore dictates a mere comprehensive approach to fertiliser utilization incorporating soil tests field research and economic evaluation of results. This is the economic relevance of soil test calibration. The biological relevance of soil test calibration is to establish the relationship between the level of soil nutrients as determined in the laboratory and the crop response to fertilization observed in the field. Such relationship permits balanced nutrition of crops. Paper focus on advantages of soil testing, soil testing service in India and Jammu and Kashmir, approaches in formulation of fertiliser recommendation in India, soil test based fertiliser recommendations, critical limits, target yield concept, fertiliser adjustment equation, site specific nutrient management, integrated plant nutrient supply system, diagnosis and recommendations and conclusions.

6.1 SOIL TESTING SERVICES IN INDIA

Soil testing programme in India starts in 1955–56 with the setting of 16 soil testing laboratories under Indo-US Cooperation Agreement on 'Determination of Soil Fertility and Fertiliser Use'. With the passage of time and modernisation of agriculture the STLs (Soil Testing Labs) increases and at present there are 514 STLs with analysing capacity of 6.4 million samples and percentage utilization of 76 per cent[1]. A brief on soil testing service are given in Table 6.1.

Table 6.1 Soil testing services in India

Total no. of soil testing laboratory			Analysing capacity (annual)	No. of samples analysed	Percentage utilization
Mobile	Static	Total			
118	396	514	6422200	4841093	75.38

Source: (1)

6.2 ADVANTAGES OF SOIL TESTING IN RELEVANCE TO AGRICULTURE

1. To sort out deficient areas from non-deficient ones
2. To help in understanding the inherent fertility status of the soils
3. Recommendations of fertilisers to be applied
4. Best guide for efficient use of fertilisers
5. Classify soils on the basis of nutrient status as low, medium and high
6. Reclamation of problematic types of soils

6.3 APPROACHES IN FORMULATION OF FERTILISER RECOMMENDATIONS

Since Liebigs time around 1840, many methods and approaches have been tried to get a precise or workable basis for predicting the fertiliser requirement of crops. They are:

1. Generalized recommendation: Optimum doses of fertiliser are obtained through various experiments, e.g. hybrid rice 120:60:40 NPK kg/ha
2. Fertiliser recommendations: Based on soil fertility categories of low, medium and high
3. Soil test based fertiliser recommendation for a certain percentage of yield maximum. Mitscherhich and Bray equation, i.e. Log $(A\text{-}Y)$ = Log $A\text{-}Cb\text{-}Cx)$
4. Fertiliser recommendation based on soil critical limits
5. Fertiliser recommendation based on target yield approach of STCR
6. Site-specific nutrient management (SSNM) for specific nutrients in crop
7. Diagnosis and recommendations system (DRIS)

6.3.1 Generalized Recommendations

This is based on fertiliser rate experiments conducted at many locations by various agencies. From these results on optimum dose of fertiliser is recommended for a crop in a given agro-climatic regions. Recommendations such as 120:60:40 kg/ha of N, P_2O_5 and K_2O, respectively is an example of this approach for high yielding varieties of rice.

6.3.2 Fertiliser Recommendations Based on Soil Fertility Categories

The soil tests are calibrated into different fertility categories, such as low, medium and high. An example of such classification for field crops is given below. The general fertiliser recommendations are equated to the soils rated as medium. For soils testing low or high category the fertiliser recommendation is increased or decreased by 30–50 per cent of general recommendations in case of field crops. The rating chart for organic carbon and available N, P and K is given in Table 6.2.

Table 6.2 Rating chart for organic carbon and available N, P and K.

S. No.	Nutrient	Low	Medium	High
1.	Organic carbon	Below 0.5%	0.5–0.75%	>0.75%
2.	Available N kg/ha	<280	280–560	>560
3.	Available P kg/ha	C10	10–25	>25
4.	Available K kg/ha	<108	108–280	>280

Source: (2)

6.3.3 STCR for a Certain Percentage of Yields

The complexity of STCR studies arises due to the diversity of soils, climate, crops and management practices. Inspite of these complexities, there are many successful attempts by scientists in establishing relationship between yield function and soil test values (STV) and making soil testing as a base for STCR in recommendations of fertilisers[6]. This is commonly known Mitscherlich and Bray approach. In this an empirical relationship is developed between percent yield and soil and fertiliser nutrient. So that fertiliser doses can be recommended for various percentage of maximum yield.

The Mitscherlich and Bray equation is: $\text{Log}(A\text{-}Y) = \text{Log}(A\text{-}Cb\text{-}Cx)$

where,

A = maximum yield
Y = percentage yield
C = proportionality factor
b = soil test value
x = dose of fertiliser added

6.3.4 Fertiliser Recommendations Based on Soil Critical Limits

Soil test are also calibrated from the data of multiplication fertiliser rate trails to deduce a critical limit for STV below which there will be positive response to added fertiliser and above which there is negative response which becomes based for recommendation of fertilisers[4].

6.3.4.1 Critical limits

It is limit of a plant nutrient below which plant started to showing deficiency. The recommended critical limits of soil available zinc for some field crops (DTPA extraction procedure) are given below here for references. The recommended critical limits of soil available zinc for some field crops are given in Table 6.3.

Table 6.3 Recommended critical limits of soil available zinc for some field crops

S. No.	Crop	Soil order	Critical limit
1.	Rice	Vertisols	1.00
		Alfisols	0.70
		Mollisols	0.80
		Inceptisols	0.70
2.	Wheat	Alfisols	0.60
		Entisols	0.50
3.	Maize	Alfisols	0.50
		Entisols	0.70
4.	Mustard	Entisols	0.50
		Inceptisols	0.50

Source: (7)

6.3.5 STCR for Target Yield of Crops

The Liebig's law of minimum states that the growth of plants is limited by the plant nutrient element in the smallest quantity, all other being present in adequate amounts. Ramamoorthy *et al.*[10] established the critical basis and experimental proof for the fact the Liebig's law of minimum operates equally well for N, P and K. This forms the base for fertiliser application for targeted yield, advocated by Troug[19].

6.3.5.1 Soil test crop response (STCR) for correlation approach

STCR approach was put forth by Trough[19] and was further modified by Ramamoorthy *et al.*[10] This approach was used for meaningful fertiliser recommendations by taking into consider the following parameters:

- Nutrient requirement
- Efficacy of fertiliser market
- Contribution from soil nutrient

These parameters are used to formulate the fertiliser prescription equations for the target yield of cropping system.

6.3.5.2 Objectives

- To establish a relationship between soil test and crop response to fertilisers on representative soils in different soil and agronomic regions of the country and the results so obtained, to provide a basis for fertiliser recommendations for maximum profit/ha
- To derive a basis for fat recommendation for desired targets suited to the constraints of fertiliser availability or credit facilities to the farmer
- To devise a burn for fertiliser recommendation for a whole cropping system based on initial soil test values

6.3.5.3 Uses of STCR

- Fertiliser prescription equations along with calibration charts are used for achieving target yields of different crops in different states of our country
- Allows balanced use of fertiliser under resource constraints and maintenance of soil fertility
- The applicability or use of this approach results in higher response ratio over general recommended doses
- Eco-friendly in nature

6.3.6 Target Yield Concept and Adjustment Equation

Ramamoorthy *et al.*[10] established the theoretical basis and experimental proof for the fact that Liebig's law of minimum operates equally well for N, P and K. This formed the basis for fertiliser application for targeted yield. Among various methods of fertiliser recommendations, the one based on yield targeting is unique because this method not only indicates soil test based fertiliser recommendations, but also the level of yield the farmer can achieve if good agronomic practices are followed in raising the crop. Targeted yield approach also provides a scientific basis for balanced fertilization not only among the fertiliser nutrient themselves, but also the soil available nutrients. The essential basic data required for formulating fertiliser recommendation based on target yield approach are:

- Nutrient requirement in kg/qt of produce
- Percent contribution from soil available nutrient (%CS)
- Percent contribution from fertiliser added (%CF)

The basic data have been derived for various crops and soil types from the soil test crop response experiments conducted under ICAR coordination projects. On the basis of these parameters fertilisers adjustment equations were developed as follows:

$$FD = \frac{NR}{\%CF} \times T \times 100 - \frac{\%CS}{\%CF} \times STV$$

Some targeted yield equations are given in Table 6.4

Table 6.4 Target yield adjustment equations

	Basic data			Fertiliser adjustment equations
	NR	%CS	%CF	$FN = 5.46\ T - 0.32\ SN$
N	1.71	10.0	31.3	$FP_2O_5 = 2.58\ T - 2.67\ SP$
P_2O_5	0.48	22.4	19.2	$FK_2O = 2.82\ T - 0.68\ SK$
K_2O	2.96	59.1	104.9	

Soil available nutrient (kg ha^{-1})			Fertiliser nutrient required for yield target (45 q/ha)		
N	P	K	N	P_2O_5	K_2O
150	5	50	198	103	93
200	10	75	182	89	76
250	15	100	166	75	59
300	20	125	150	63	42

Source: (8)

6.3.6.1 Target yield equations are suitable under the following situations

- These should be used for similar soils occurring in a particular agro-ecosystem
- The maximum target should not exceed 75–80 per cent of the highest yield achieved for the crop in the area
- Adjustment equations must be used within the experimental range of soil test values and cannot be extrapolated
- Good and recommended agronomic practices need to be followed while raring crops
- Other micro and secondary nutrients should not be yield limiting

6.3.6.2 STCR Studies under integrated plant nutrient supply system (IPNS) for optimising fertiliser doses (OFDs)

The integrated nutrient management involved combined use of fertiliser; organic manure and biofertilisers are not only gaining wider recognitions, but also appears to be a great promise for the future soil fertility maintenance. This technology showed the possibility of replacing a part of non-renewable energy based fertiliser's nutrient. The increased response ratio recorded under IPNS indicated the increased use efficiency of added fertiliser nutrients, the low cost inputs and higher fertiliser use efficiency (FUE) resulted in high benefit cost ratio. Since, optimisation of fertiliser doses under IPNS assures a balanced supply of nutrients. This technology could promote soil health and productivity. An experiment was conducted at Coimbatore on rice, the basic data, fertiliser adjustment equations and doses of fertiliser nutrient for specific yield target of 50 qt/ha is shown in the following equation. The fertiliser adjustment equations development under IPNS adopting target yield concept in the soil test crop response correlation given in Table 6.5. From the results it is clear that considerable quantity of fertiliser nitrogen is reduced by the use of IPNS in combination with fertiliser nitrogen to get a desired yield target[9].

Table 6.5 Fertiliser adjustment equations development under IPNS adopting target yield concept in the STCR

Basic data

NR	%CS	%CF	CO (FYM)	CO (GM)	Co (GM + Azo)
2.28	21.6	48.30	26.90	19.90	17.20

Targeted yield equation: FN = 42.7T– 0.44 SN, FFYM = 4.72T – 0.44 SN -0.55O

$$FD = \frac{NR}{CF} \times 100\,T - \frac{\%CS}{\%CF} \times S - \frac{\%Co}{\%Cf} \times O$$

STV	**Target yield for 50 q ha^{-1}**			
N	FN	FN + FFYM	FN + GM	FN + GM + Azo
250	126	91 + 35	81 + 45	70 + 48
275	115	80 + 35	70 + 45	67 + 48
300	104	69+ 35	59 + 45	57 + 48

Source: (7)

6.3.6.3 Effect of IPNS system vs fertilizer alone on crop response

The response of rice to various IPNS as compared to farmers practice and STCR without combination of IPNS for the same target yield were found to be high. The higher response under IPNS indicated that under high input production system, where productivity cannot be further increased with increased use of mineral fertilisers alone. IPNS could increase higher use efficiency of fertiliser added[11]. The higher response ratio was also recorded. In STCR technology the fertiliser doses are tailored to the levels of crop taking into account the soil contribution and the contributions from components of IPNS. So, there is a balanced supply of required quantity of nutrients to the crop and thereby wastages are avoided. This prevents pollution of soil and paves way for higher returns. The higher BCR recorded for rice under IPNS clearly indicates the possibility of getting better profit under IPNS than the use of mineral fertiliser alone. The effect of IPNS system vs fertiliser alone on crop response is given in Table 6.6.

Table 6.6 Effect of IPNS system vs fertiliser alone on crop response

S. No.	Treatment	Quantity of nutrient added			Yield t/ha	Response over farmers practice (kg/ha)	Response ratio (kg/kg)
		N	P	K			
1.	Farmers practice	58	–	293	21.5 or 2/500 kg/ha	–	–
2.	STCR-40 t/ha yield target	167	132	256	38 or 38000 kg/ha	16500 kg/ha	29.7
3.	STCR-40 t/ha yield target under IPNS (FYM) @ 12.5 t/ha	83	113	210	40.2 or 40200 kg/ha	18700 kg/ha	46.1
Soil test values kg/ha =		N	P	K			
		185	11.0	180			

Source: (5)

The STCR tell us which crop has more nutrient uptake capacity than others. For this an experiment was conducted by Milap Chand *et al.*[8] in Ludhiana on rapeseed and mustard. Based on seed yield data, uptake of nutrients and STV and applied fertiliser nutrients, the estimates of nutrient requirement (kg/t), %CS and %CF for N, P and K have been developed. The results showed that rapeseed had higher N, P and K requirement than mustard for unit production of seed. The capacity of mustard and rapeseed to exploit soil available pool of N, P and K was similar. But the efficiency of rapeseed to derive nutrients from applied fertiliser was higher than mustard. Fertiliser adjustment equation for target yield of mustard and rapeseed indicated that for the same level of crop production, with the same level of soil nutrient, status, rapeseed required higher amount of N, P and K than mustard shown in the Table 6.7.

Table 6.7 Soil test fertiliser recommendation for target yield of mustard and rapeseed

Mustard			**Rapeseed**		
Soil test values kg/ha	Target yield t/ha		Soil test value kg/ha	Target yield t/ha	
	2.0	2.5		2.0	2.5
$KMnO_4$	Fertiliser N required		$KMnO_4$-N	Fertiliser N-required	
60	101	144	80	110	146
100	82	125	100	101	137
140	63	106	120	91	128
180	44	87	140	82	119
Olsens-P	Fertiliser P requirement		Olsens-P	Fertiliser P required	
4	22	32	6	22	30
8	12	22	9	18	26
12	2	12	12	14	22
16	0	3	15	11	18
Amm. Ac-K	K requirement		Amm. Ac-K	Fertiliser K requirement	
60	28	56	100	53	73
80	10	38	130	43	64
100	00	19	180	35	58

Source : (8)

6.3.6.4 Use of STCR for suitability of cropping system in various soil orders

STCR is not applicable to only one crop but is also applicable to a cropping system. It gives us reliable information that which cropping system is suitable to which type of soil orders. An experiment was conducted at different locations on the soil order inceptisols (Fig. 6.1). It was found that maize + wheat provides maximum yield that finger millet + maize and rice + rice. Another experiment was conducted on soil order alfisols, here it was found that finger millet + maize gives maximum average yield than maize + wheat and soybean + wheat (Fig. 6.2).

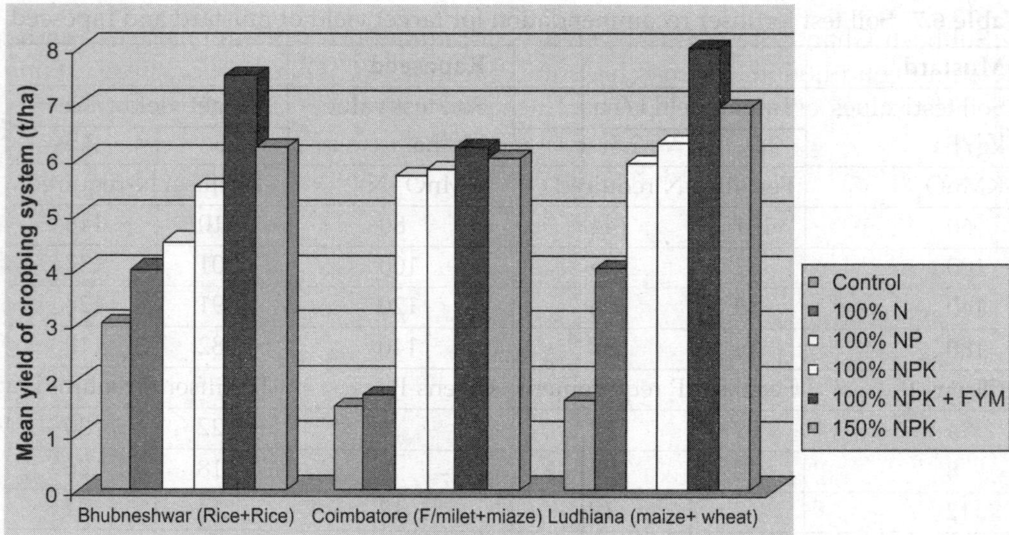

Fig. 6.1 Cropping system experiment in inceptisols order

Source: (17)

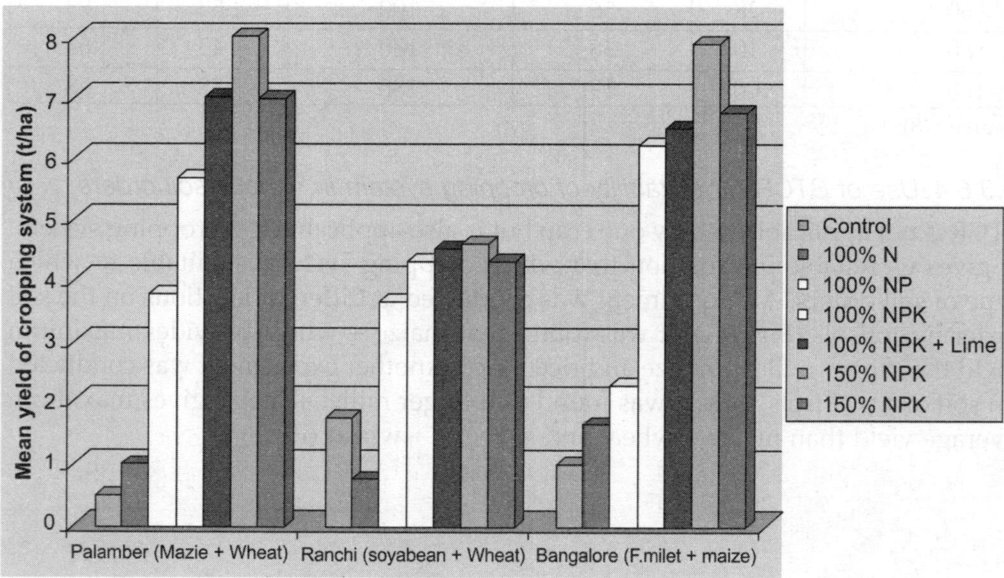

Fig. 6.2 Cropping system experiment in Alfisols*

Source: (17)

* Alfisols is a dominant soil order occupies highest soils in India next to the enseptisols and Entisols. The area of Alfisols is 43.2 M ha in India.

Subhash Chand[16] tested a targeted yield equation for Shalimar-Rice-1 under Kashmir condition in a Research Council Meeting (RCM) institutional project and recorded a highest yield of 8.5 t/ha. Integrated nutrient managements are the best ways to utilize all the nutrient sources rationally for higher crop productivity and sustainable soil health. Subhash Chand[15] shorted out more then 200 of such practices for their practical utility in sustainable and steady development farmers. Agro-ecological region wise IPNS strategies for crops and cropping systems in India are more important in decision making about uses of nutrient sources for sustainable crop productions[14, 15]. The utilisation of all the sources of plant nutrient for sustainable crop production in Indian agriculture is the need of time[11]. The biofertilizers are important component of IPNS strategies[18]. The IPNS superiority of target yield approach based on STCR over farmers practiced general recommend dose of fertiliser and recommendation of fertiliser after the soil testing (Fig. 6.3). The reasons for this are: high benefit cost ratio, eco-friendly, avoid sub or super optimal (doses of fertilisers) and high response ratio. A terminology of soil fertility, fertilizer and organics complied by Subhash Chand (2014) for further enhancing knowledge about soil science terms and definition.

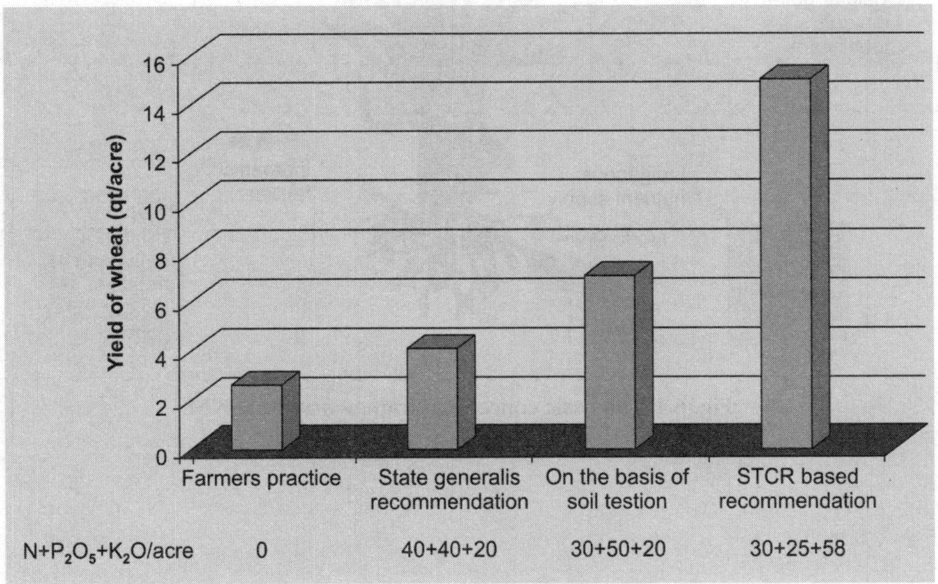

Fig. 6.3 Comparison between farmers practice, generalised recommendation, on the basis of soil testing and STCR based recommendation

Source: (17)

6.3.6.5 Site specific nutrient management (SSNM) for rice crop

Fertilisers are one of the main inputs in rice production. The quantity and management of fertilisers that best match the needs of rice crops for essential nutrients can vary greatly among fields, seasons, and years as a result of differences in crop-growing conditions, crop and soil management, and climate. Hence, the management of nutrients for rice requires an approach, which enables adjustments in applying N, P, and K to accommodate the field-specific needs of the rice crop for supplemental nutrients. Site-specific nutrient management (SSNM) in diversified cropping provides a field-specific approach for dynamically applying nutrients to cropping as and when needed. This approach advocates optimal use of indigenous nutrients originating from soil, plant residues, manures, and irrigation water. Fertilisers are then applied in a timely fashion to overcome the deficit in nutrients between the total demand by crop to achieve a yield target and the supply from indigenous sources.

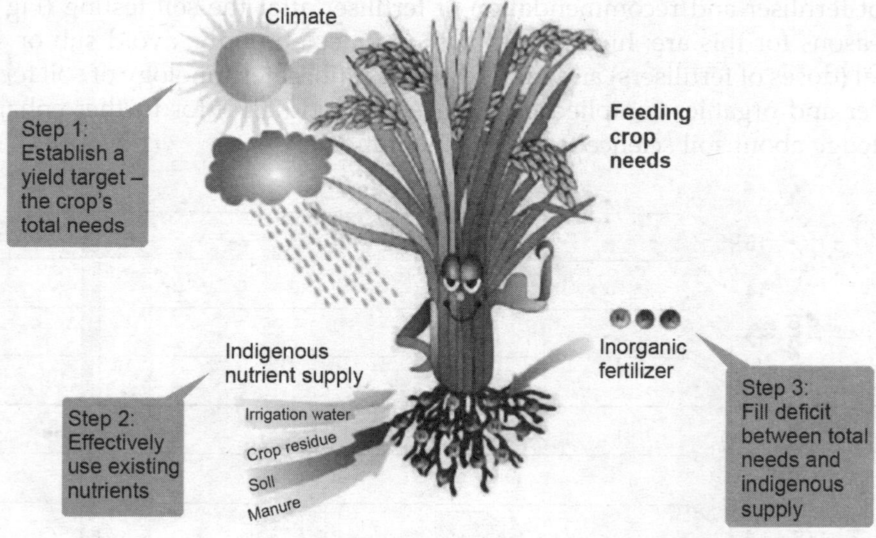

Fig. 6.4 The basic conceptual framework for SSNM

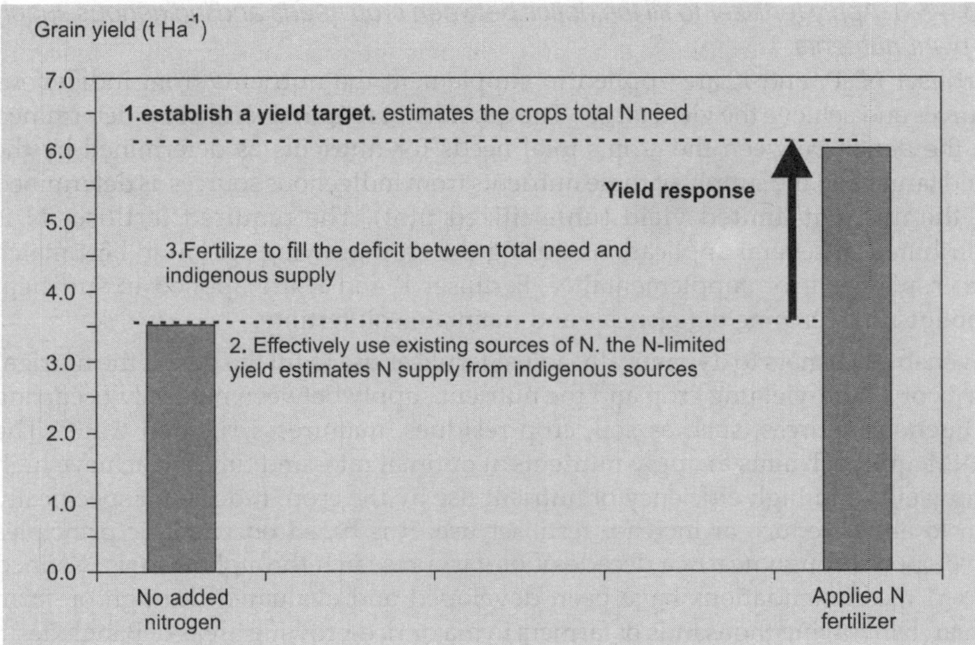

Fig. 6.5 The basic conceptual framework for SSNM

The site-specific nutrient management (SSNM) is a simple plant need-based approach for optimally applying N, P, and K to different crops. The SSNM approach involves the following three steps:

6.3.6.5.1 Establish an attainable yield target

Crop yields are location and season specific depending upon climate, crop cultivars, and crop management. The yield target for a given location and season is the estimated grain yield attainable with the farmers' crop management when N, P, and K are effectively supplied. Because the amount of a nutrient taken up by a crop is directly related to yield, the yield target indicates the total amount of the nutrient that must be taken up by the crop. The yield target typically does not exceed about 80% of climatic and genetic potential yield.

6.3.6.5.2 Effectively use existing nutrients

The SSNM approach promotes the optimal use of naturally occurring indigenous nutrients coming from the soil, organic amendments, crop residue, manure, and irrigation water. The uptake of a nutrient from indigenous sources can be estimated from the nutrient-limited yield, which is the grain yield for a crop not fertilized with the nutrient of interest but fertilized with other nutrients to ensure they do not limit yield.

6.3.6.5.3 Apply fertilizer to fill the deficit between crop needs and indigenous supply of plant nutrients

Fertiliser N, P, and K are applied to supplement the nutrients from indigenous sources and achieve the yield target. The quantity of required fertiliser is determined by the deficit between the crop's total needs for nutrients as determined by the yield target and the supply of these nutrients from indigenous sources as determined by the nutrient-limited yield (unfertilized plot). The required fertiliser N is distributed in several applications during the crop growing season to best match the crop's need for supplemental N. Fertiliser P and K are applied in sufficient amounts to overcome deficiencies and maintain soil fertility.

It enables farmers to dynamically adjust fertiliser use to fill the deficit the nutrient needs of a high-yielding crop and the nutrient supply between naturally occurring indigenous sources, such as soil, crop residues, manures, irrigation water. The SSNM approach aims to apply nutrients at optimal rates and times to achieve high crop yields and high efficiency of nutrient use by the crop. It does not specifically aim to either reduce or increase fertiliser use. It is based on scientific principles developed through nearly a decade of on-farm research throughout Asia. Refined SSNM recommendations have been developed and evaluated through on-farm research involving thousands of farmers in major rice-growing areas of Bangladesh, China, India, Indonesia, Myanmar, the Philippines, Thailand, and Vietnam. The key features of the SSNM approach in diversified cropping include:

- Dynamic adjustments in fertiliser N, P, and K management. It should be field- and season-specific according to the crops grown in field;
- Effective use of indigenous nutrients according to crops grown in field;
- Fertiliser N management through the use of the leaf color chart (LCC), which helps ensure N application with respect to time and amount needed by the crop;
- Use of the omission plot technique to determine the requirements for P and K fertiliser;
- Managing fertiliser P and K to both overcome P and K deficiencies and avoid the mining of these nutrients from the soil

The total amount of required fertiliser N can be approximated from the anticipated crop response to fertiliser N application, which is the deference between attainable target yield and N-limited yield (i.e. yield with no fertiliser N and no limitation of other nutrients). The estimated total fertiliser N requirement by the crop is then apportioned among multiple times of application during the growing season to ensure that the supply of N matches the crop need at critical growth stages. A key ingredient for dynamic N management is a method for the rapid assessment of leaf N content, which is closely related to photosynthetic rate and biomass production and is a sensitive indicator of changes in crop N demand within a growing season. The LCC is an inexpensive and simple tool for monitoring the relative greenness of a rice leaf as an indicator of leaf N status, and then enabling farmers to apply fertiliser N whenever leaves reach a critical N status determined by their yellowish-green color.

The specific objectives of the SSNM in diversified cropping are to obtain target yield from existing cropping system with maximum profit and to maintain sustainability of the soil. It includes nutrient management according to site (field to field), crop-to-crop, crop-rotation to crop-rotation, season to season. SSNM strives to enable farmers to dynamically adjust fertilisers use to optimally fill the deficit between nutrient need of a high yielding crop (s) in a system and the nutrient supply from naturally occurring sources. Total nutrient management helps in increasing nutrient use efficiency (NUE) and saves fertilisers.

6.3.7 Diagnosis and Recommendations Integrated System (DRIS)

The dynamic system nature of leaf composition which is governed by a large number of inherent factors, environment and management makes foliar diagnosis a complex exercise. A process involving multiple steps has been worked out and is known as DRIS. The DRIS was developed by Beaufils (3) and used by several workers. DRIS is a holistic system of nutrient diagnosis and can be used for isolating nutrients and other factors which determine yield level and quality of produce. The system is capable of following:

- Making diagnosis of index tissues with variable ages
- Identifying deficiency or toxicity of nutrient
- Predicting the probably suboptimal status of an element, on correcting the most deficiency element
- Listing the nutrient elements in order of their limiting importance on yield

The DRIS was used to develop norms of leaf nutrients and soil fertility for sustainable crop production. It has advantage over other methods as it allows index tissue sampling for wider period of time, the diagnosis is based on large number of nutrients.

6.4 POSITION AND FUTURE LINE OF WORK

- Calibration of time tests as adjunct to soil testing
- Role of sub-soil contribution in soil test calibrations
- Development of soil test calibration for variations in management, sowing time, etc.
- Soil, water and plant testing labs should be strengthened in term of infrastructure and sophisticated equipments
- A computer aided data base should be created for all the nutrients, sources of nutrients, their compositions, transformations and release, fixation of nutrient and energy turnover of nutrient sources
- A database on integrated nutrient management for sustaining crop productivity and soil health should be created in recommendations form for ready uses by farmers and extension practitioners
- Awareness need to be created among farmers through farmers trainings, field visits crop seminars, *kisan melas* and field demonstrations about soil health and nutrient management for sustainable crop production.

6.5 CONCLUSIONS

Balanced use is the key to increase crop productivity. STCR approach is the best method to know the nutrient imbalances in the soil and apply required amount of nutrients to optimize crop production as well as crop nutrient. The method STCR is eco-friendly with high response ratio and adoptable technology to be used by developing countries. However, STCR using IPNS is a more practical approach in sustainable crop and soil management. The DRIS is a right approach for finding out the right doses for foliar concentration and doses for quality fruit production.

6.6 KEY REFERENCES AND RESOURCES FOR FURTHER READING

1. Anonymous, Soil Testing Services in India and J&K. Ministry of Agriculture. Department of Agric. and Coop., Krishi Bhavan, New Delhi 2002.
2. Baver A, Natesan R. *Fert News* 2001; 20(4):64–70.
3. Beaufils ER. *Fert Svc South African J* 1971; 1:1–30.
4. Cate R.B Jr, Nelson LA. North Carolina Agric. Exp. Stn. Int. Soil Testing Series *Tech. Bull.* No. 1. (1965)
5. Kadam BS, Sonar KR. *J Indian Soc of Soil Sci* 2006; 54(2):513–515.
6. Kanwar JS. Soil Testing Service in India Retrospect and Prospect. In proc. of Int. Sym. on soil Fertility Evaluation, New Delhi 1971; 1:1103–1133.
7. Katyal JC. *Fert News* 1985; 30(4):6–80.
8. Milap Chand DK, Benbi and AS Azaad. *J Indian Soc of Soil Sci* 2006; 258–261.
9. Palaniappan SP, Prasad TVR. In Resource Management for Sustainable Crop Production 1999; 280.
10. Ramamoorthy B, Narsimhan RL, Dinesh RS. *Indian Farming* 1967; 25(5):43.
11. Rajendra Prasad. *Indian J Fert* 2008; (4)12:65–70.
12. Roy RN. In Fertilisers, Organic Manures Recycling Wastes and Biofertilisers and Components of Integrated Plant Nutrition. Tandon, HLS, (Ed.). FDCO, New Delhi (1999).
13. Selva Kumari G, Santhi R, Mathan KK. *Bulletin of IISS* 1999; 101.
14. Subba Rao A, Chand Subhash, Srivastava S. Fert News 2002; 42:75–95.
15. Chand Subhash (Ed.). Integrated Nutrient Management for Sustaining Crop Productivity and Soil Health, I. B. D. Co. Lucknow 2008; p112.
16. Chand Subhash (PI). Concluded Project on Testing of Validity of Fertiliser Prescription Equation for Rice under Shalimar Condition. Division of Soil Science, SKUAST-K 2008; 13.
17. Swarup A, Srinavasa Rao. Fert News 1999; 30(2):67–81.
18. Tewatia RK, Kawle SP, Chaudhary RS. *Indian J Fert* 2008; 3:1:111–118.
19. Troug E. In Fifty Years of Soil Testing. Transac of 7th Int. Congress of soil sci, Madison Wisconsin, USA, Part III and IV 36–45 (1960).
20. Chand S. Terminology of Soil Fertility, Fertilisers and Organics. Astral International, 2014; p179.

Fertilisers are the cheapest source of plant nutrients. Use fertilisers wisely and economically for higher production and productivity. — Subhash Chand

CHAPTER 7

Integrated Nutrient Management (INM) Options for Sustainable Agriculture and Social Security

Science without religion is lame, religion without science is blind.

—Albert Einstein

Indian agriculture is witness to several crucial issues like low fertilizer use efficiencies, loss of soil biodiversity, adverse effect of climate change (drought, tsunamis, earthquake and landslides), decreasing agriculture growth, decreasing factor productivity, low soil organic carbon (SOC) stock, mismatch between nutrient removal and addition to the soil (10–12 mt of nutrient gap), disparities in the NPK ratio, emergence of multiple nutrient deficiencies (B, Zn, Cu, S) and water scarcity. The whole scenario of agriculture is at a junction where one has to rethink and reform the agricultural packages and practices to fulfill the dreams of million peoples of the country. All these challenges are resultant of green revolution by which we have attained self sufficiency in food grains production. Now in the next 50 years, the challenges for agriculture will not only be to meet the food needs of the world expanding population, but also undertake it in a manner that is sustainable and profitable for future generation. The fertilizers are very important sources of plant nutrients and played a prominent role in increasing food grain production of the country. About fifty percent increase in the food grain production in post green revolution era is attributed to the use of fertilizers. Soil is a precious natural resource equally as important as water and air. The proper use of soil greatly determines the capability of a life-support system and the socio-economic development of a country. The agriculture era has been changed from resource degrading to resource conserving technologies and practices which will enable help for increasing crop productivity besides maintaining soil health for future generations. The INM provides an excellent opportunity not only for sustaining soil, but enhancing crop productivity also. The INM is the maintenance or adjustment of soil fertility and plant nutrient supply to an optimum level for sustaining the desired crop production through optimization of the benefits from all possible sources of plant nutrients in an integrated manner. INM options proved to be helpful in food, nutritional and livelihood security besides socio-economic conditions of Indian farmers. This articles describe all the possible INM option for rice, wheat, sorghum, maize, bajra, soyabean, groundnut, sunflower, cotton, mustard, sugarcane, pulses, vegetables, spices, fruit crops, ornamental plants and miscellaneous crops. The INM options also given to various cropping systems, viz. rice-wheat, rice-rice, maize-wheat, soyabean-wheat and miscellaneous cropping systems. The article provides onetime solution for sustainable soil, crop and management of Indian agriculture.

7.1 CONCEPTS OF INM

The continuous and imbalance use of fertilizers is adversely affecting the sustainability of agricultural production besides causing environmental pollution. The major issue for the sustainable agricultural production will be management of soil organic carbon and rational use of organic inputs, such as animal manure, crop residues, green manure, sewage sludge and wastes known as integrated plant nutrient resource management. However, since organic manure cannot meet the total nutrient needs of modern agriculture, hence integrated use of nutrients from fertilizers and organic sources will be the need of the time. The possible reasons for the apparent decline in returns from the increased fertilizer applications include:

- Imbalanced N, P and K application and the latter two are being applied in too low amounts in some states
- Deficiencies of secondary and micronutrients are appearing with increasing frequency
- More fertilizers are being used on soils inherently poorer in fertility and/or uncertain water supply as in dryland areas
- The increased intensity of cropping together with changes in crop sequences, e.g. cereal-cereal rotations in place of cereal-legume rotations
- There may be an overall decrease in soil organic matter status of soils
- Built up of certain diseases, pests and weeds under intensive system of cropping
- Use of fertilizers, nutrients in high amounts in some intensively cultivated areas may lead to serious problems of deterioration of the soil quality as a result

To overcome the negative effects of application of plant nutrients, both at low or imbalanced and high levels of input can be avoided by excellent management. Balanced fertilization supplemented with organic nutrient sources help in overcoming the hazards of nutrient depletion and of mining soil fertility. Integrated nutrient management (INM) provides excellent opportunities to overcome all the imbalances besides sustaining soil health and enhancing crop production. The concept of INM is the maintenance or adjustment of soil fertility and of plant nutrient supply to an optimum level for sustaining the desired crop production through optimization of the benefits from all possible sources of plant nutrients in an integrated manner. The INM as defined by Harmsen here differs from the conventional nutrient management by more explicity considering nutrient from different sources, notably organic materials, nutrients carried over from previous cropping seasons, the dynamics, transformations and interactions of nutrients in soil, interaction between nutrients, their availability in the rooting zone and during growing season in relation to the nutrient demand by the crop. In addition it integrates the objectives of production, ecology, environmental and is an important part of any sustainable agricultural production system. The objectives of INM may be classified as short objectives and broad objectives.

7.1.1 Shorter Objectives of INM

To maintain or enhance soil productivity through a balanced use of mineral fertilizers combined with organic and biological sources of plant nutrients. To improve the stock of plant nutrients in the soil. To improve the efficiency of plant nutrients, thus limiting losses to the environment.

7.1.2 Broader Objectives/Goal of INM

To increase the availability of nutrients from all sources in the soil during the growing season. To match the demand of nutrients by the crop and supply of nutrients from all sources through the labile soil nutrient pool, both in space (the rooting zone) and time (the growing season). To optimize the functioning of the soil biosphere with respect to specified functions, such as the decomposition of organic matter (mineralization), the control of the pathogenic organisms by their natural enemies (predators), the biological formation of soil structure (aggregates, biopores), the decomposition of phytotoxic compounds, etc. To minimize the losses of nutrients to the environment, e.g. through ammonia volatization and denitrification in the case of nitrogen, surface run-off and leaching beyond the rooting zone.

The INM is a prescription for excellent soil health. Increasing sustainability concerns and shortage of fertilizers require a system of nutrient management that makes optimum use of nutrients from soil resources. Efficient recycling and use of organic as a carrier of nutrients, synergistic and antagonistic effect of nutrients on plant nutrition in cropping system for purpose of developing appropriate fertilizer plans. In this context INM is necessary and we need to put it in place. The IPNS system is a "System Approach" to farming, possibly it should have four major components, i.e. (a) on-site nutrient resource generation, (b) mobilisation of off-site nutrient resources, (c) resource integration, and (d) resource management. In recent years a lot of emphasis has given for integrated use of inorganic fertilizers and organic manure have become important for higher agricultural production. No single source of plant nutrients, be it chemical fertilizers, organic manure, crop residues, green manure or even-biofertilizers can meet the entire nutrient needs of crops in modern agriculture.

7.2 COMPONENTS OF INM

7.2.1 Soil Reserves (SRs)

Soils considered as a mother for nurturing plants and supply all the 16 essential and 4 beneficial plant nutrients. Nutrients are mostly found in organic and/or fixed mineral form. Plants can meet much of their nutritional requirement form this source, if managed properly, mainly through mineralization of soil organic matter. But due to continuous and intensive cultivation, the nutrient supplying capacity of soils has been decreased considerably. Therefore, under any intensive agriculture system, special emphasis should be given to raising SOM to maintain soil nutrient and to reduce soil degradation. To enhance soil nutrient supply it is necessary to adopt appropriate soil management practices, such as improvement of soil physical conditions and addition of appropriate quantities of nutrients including micronutrients through mineral fertilizer, organic and biological sources.

7.2.2 Use of Mineral Fertilizers

Fertilizers are industrially and chemically manufactured source of plant nutrients. Various types and grades of fertilizer are available throughout the country for supplying major nutrients such, as N, P, K and micronutrients. The fertilizer use levels differ widely between various soils and nutrient use is mostly imbalanced, favoring excess use of nitrogen. Balanced fertilization is known to improve fertilizer use efficiency (FUE) and at the same time profitability for the farmer. Using ever higher rates of nitrogen (urea mostly) alone with improved better varieties, the resulting higher yields also remove ever larger amounts of soil nutrients if not replenished and the FUE declines further resulting in stagnating and even declining yields. This leads to the paradox situation where statistics report the continuing increase in fertilizer use, but the expected crop production increases are not taking place, it is also known as yield plateau. Apart form N, P, K and S and micronutrients, such as zinc (Zn), iron (Fe), manganese (Mn) and boron (B) have also gained importance in recent years. The secondary nutrient sulphur has become deficient over wide areas especially since the intensive use of high analysis fertilizer, urea, instead of sulphate of ammonia and TSP or DAP instead of single superphosphate or NPK compounds. To overcome the imbalances of nutrients, it is essential to supply a part of nutrient demand through the use of organic sources which will result in good soil biodiversity besides improvement in soil bio-physico-chemical properties in soil systems.

7.2.3 Organic Sources of Plant Nutrients

The sustainability of highly intensive cropping systems and the associated heavy mineral fertilizer use without organic manures is widely questioned. This has brought the almost forgotten farmyard manures (FYM) and composts back to the forefront. Regular applications of such organic manures not only supply all the various secondary and micronutrients, though in small quantities, but also improve the physical and biological properties of the soil. Furthermore, return to the farm is the best way to take care of the large amounts of animal waste produced in the commercial animal husbandry, pig and poultry farms, instead of dumping and degrading the environment.

7.2.3.1 Farm yard manure (FYM)

FYM is a very good source of plant nutrients, however, traditionally does not receive the attention it deserves, as most farmers store their most valuable asset, their cattle/buffalo manure not in a systematic, but in a rather haphazard way. Storage of FYM in rural households in the country is in heaps exposed to sun, wing and rain, which accounts for substantial nutrient losses. FYM preparation needs improvement, adhering to strict and prompt coverage for shading and prevention of drying out by hot wind or washing out of nutrients with heavy rains (pollution hazard).

7.2.3.2 Composting

Compost is not a by-product of common farm activities, but has to be specially prepared for its own sake. The quality of the ripe compost after undergoing a heating process reaching at least 60 °C to destroy harmful pathogens and weed seeds will depend on the raw material used and the attention given to proper composting by the farmer. The C – N ratio needs to be lowered to 20–15 and good quality compost should have no more than 30 per cent moisture, as no farmer wants to carry excess water to the field. Practically all sixteen plant nutrients are contained in compost, but in very small quantities. For phospho-compost preparation, it is suggested to mix 3 per cent of SSP with composting materials.

Composting is a labor and time-consuming process, which takes 3–6 months. To speed up the process in several countries, rapid composting technologies have been developed. With the use of *Trichoderma harzianium*, a fungal activator, decomposition of rice straw and other organic material with high C – N ratio, combined with animal manure is enhanced to 25 days. Commercially prepared composts marketed as organic fertilizer are available in the countries and used mainly for high quality vegetable production and horticultural use. Several private companies coming forward for manufacturing compost for farmers' use.

7.2.3.3 Crop residues management (CRM)

Crop residues are freely available sources of organic matter that available on — farm in large quantities are wheat and rice straws, maize stalks, and stovers of legumes and various pulses. Most of the crop residues are not collected for composting and nutrient recycling, but are used as animal feed (straws/stovers), burnt or left in the field for natural decomposition (fallen leaves and stubble). Crop residues in the long run also increase the OM content in the soil. Mulching with flesh straw or leaves is another good agronomic practice for conserving moisture, reducing soil erosion and for recycling of nutrients, if the partly decomposed mulching material is ploughed under for the following crop. Direct seeding of maize or soybean into mulch cover would be another good agronomic practice. Burning of straw which is still widely practiced by farmers as the fastest and least labor requiring method of disposal should be discouraged.

7.2.3.4 Green manuring (GM)

Green manure crops such as *Sesbania aculeate* ploughed into the soil after 45–60 days may contribute about 30–40 kg/ha nitrogen for the following crop. However, due to limitations to grow a green manure crop, which occupies the land for several months and needs water and fertilizer, except N-just to plough it back into the soil. Wherever possible and feasible the growing of grain legumes, such as groundnuts, soybeans, chickpeas, cowpeas or mungbean as cash crops, which maintain soil fertility and provide farmers with extra income and fodder from crop residues should be encouraged for improving soil physico-bio-chemical properties.

7.2.3.5 Biogas slurries (BgS)

Biogas plants in rural areas produce digested slurry as an end product, which could be applied directly in cultivated fields. Such slurry contains about 1.5–2.0 per cent nitrogen, 1.0 per cent phosphorus and a little over 1 per cent potassium. It is also a valuable source of micronutrients. Moreover, due to the heated digestion processes, biogas slurry is virtually free from weed seeds and pathogens.

7.2.3.6 Industrial waste materials (IWM)

The most industrial waste materials are valuable resources and should be properly managed and utilized. The large number of sugarcane processing factories in the region produces substantial quantities of organic by-products such as biogases, pith and press-mud. Even though some of the biogases and came residues are burnt as fuel in the sugar industry. So far only a small portion is mixed with press-mud, composted and recycled as organic fertilizer. Agro-industries, such as fruit and vegetable processing, cotton ginneries, oil mills, breweries and distilleries, also produce large quantities of organic waste materials which need to be properly managed and utilized for nutrient recycling instead of dumping and polluting the environment.

7.2.3.7 Cit refuse (Garbage, sewage sludge)

Increasing population and even faster growth of urban population will consequently lead to increasing amounts of urban waste, which would create enormous disposal problems if not properly recycled as a source of crop nutrients. Processed, composted solid organic wastes and sewage sludge provide both organic matter and valuable plant nutrients to crops. The transport from urban composting plants to the farming areas constitutes a major part of the cost of processed organic wastes for farmers. Marketing studies and advertising campaigns, attractive comparative prices together with a subsidy scheme to encourage the large-scale acceptance by farmers of urban compost should be considered. Subsidies, grants and credit should concentrate on transport and handling cost of such bulky products, which could nevertheless result in considerable saving in mineral fertilizer, for the farmers. As a rule of thumb the price per kg of nutrient in composted city refuse for the farmer should be at par or not considerably higher than the cost per kg nutrient in commonly used mineral fertilizer. The other not so easily quantifiable benefits of using organic fertilizer materials, such as increasing SOM, better water holding capacity, and better soil health, are to be accounted for by the cost of extra labor for spreading and incorporation in the field.

7.2.3.8 Enriched city compost (ECC)

City compost produced at mechanical composting plants throughout the country is generally low in plant nutrients and therefore its acceptability by farmers has been limited. To improve the quality and nutrient content of city compost low-grade rock phosphate and phosphate solubilising *Azotobacter* spp. and the nitrogen fixing bacteria, such as *Azotobactor* spp. or *Pseudomonas* spp. are being used as inoculants. Microbial inoculation and application of 1 to 5 per cent rock phosphate increased the nitrogen content of city compost by 24 to 30 per cent and more favorable C — N ratios have been obtained. Available P_2O_5 content of compost was increased by 60 to 114 per cent where rock phosphate was applied and inoculated with *Aspergillus awamori*. Preparation of compost from enriched city garbage or otherwise is promosing, provided that financial support from government is available. However, heavy metals in sewage sludge when continuously applied in excessive quantities to farmland as organic manure could lead to problems. Monitoring for Cd, Zn, Pb, As, and Cu contents in compost is recommended.

7.2.4 Biofertilisers — Sources of Plant Nutrition

Biofertilizers have an important role to play in rainfed areas in improving the nutrient content of the crop. Although *Rhizobium* is the most researched and well known among the biofertilizers, there are a number of microbial inoculants with potential practical application in INM. Such inoculates could contribute to increasing crop productivity through increased biological nitrogen fixation (BNF), increased availability or uptake of nutrients through phosphate solubilization, or increased absorption, stimulation of plant growth (hormones), or by rapid decomposition of organic residues (rapid composting technology).

7.2.4.1 Rhizobium inoculants

The nitrogen fixed by rhizobia benefits legume crop production in two ways a. by meeting most of the legume crops nitrogen needs, and b. by enriching the soil for the benefit of subsequent crops. *Rhizobium* inoculation should be considered in all legume green manure crops to gain maximum benefit from nitrogen fixation in the shortest possible time. *Azospirillum, Azotobacter* and *Pseudomonas* inoculations on upland rain crops are still in their infancy and field trial results are inconclusive, although good responses to *Azospirillum* and *Azotobacter* inoculation of wheat, rice and sugarcane have been recorded.

7.2.4.2 Biofertilizers for flooded rice

Blue green algae (BGA) and *Azolla* are most important biofertilizers for wetland rice are the water fern *Azolla* and the blue green algae (BGA), also known as floating nitrogen fertilizer factories. Both can grow alongside paddy. *Azolla* can also be used for green manuring which could contribute from 20 to 60 kg per hectare N. Phosphorus is a key element and its deficiency results in poor growth and reduced N fixation (addition of 1 kg P results in fixation of 5 to 10 kg N). *Azolla* is considered an efficient scavenger for potassium and serves as a source of K for rice. Azolla biofertilizer technology is labor intensive. Irrigation water, phosphate fertilizer and pest control measures are necessary inputs. Nitrogen fixed by azolla or BGA becomes available to the rice crop only after its decomposition. Numerous field experiments indicated that only up to one third of the fixed N is absorbed by the following rice crop, while two-thirds remained in the soil as residual nitrogen or is lost to the atmosphere.

7.2.4.3 Phosphate solubilizing microorganisms (PSMs)

A variety of bacteria and fungi has been identified to have the ability to solubilize and transform inorganic P from normally insoluble sources through excretion of various organic acids have been isolated. These are bacteria of the *Bacillus* and *Pseudomonas* spp. and fungi, such as *Aspergillus, Penicillium* and *Trichoderma* spp. In addition to p-solubilization these microorganisms can also mineralize locked up organic P into soluble, plant available forms. As these reactions take place in the rhizosphere and the microorganisms bring more P into solution than they can absorb for their own growth, the surplus is available for plants to absorb. The effectiveness of these microorganisms depends on the availability of sufficient energy sources like carbon in the soil phosphorus concentration, particle size of rock phosphate as well as temperature and moisture.

7.2.4.4 Constraints to use Biofertilizers

As a result of efforts made by various agencies during the last over one decade together with Government support, there had been considerable improvement in the popularity of biofertilizers. However, the achievement is far from satisfactory. This is simply because the crop responses to biofertilizers like organic manures are not as instant and spectacular as is possible with the use of chemical fertilizers. Also these are not available as readily as the chemical fertilizers. Thus they suffer from these and score of other constraints which are highlighted hereunder:

7.2.4.4.1 Production constraints

- **Raw material:** Peat is regarded as an ideal carrier for biofertilizers. However, it is not available in India in sufficient quantities and in desirable quality. The other suitable carrier is lignite. That too is not available readily and is quite costly. In addition, materials like charcoal, press mud, agro-industrial wastages, compost, etc. have also been tried. In tropics, agro-industrial wastes may prove better alternatives to peat, etc. provided proliferation of contaminants is checked by a suitable anti-microbial agent, but for the cost. Storage of carrier based inoculants has also been found that were prepared in a mixture of soil and charcoal or press mud and charcoal.

- **Specificity of strains:** Most of the biological strains of biofertilisers are soil and agro-logical strains specific. This limits their wide spread and fool-proof use with expected performance. Therefore, region wise inadequate availability of specific strains limits their popular use.
- **Biological constraints:** Presence of ineffective or antagonistic strains in the bio-inoculants, which cannot be displaced easily, mars the overall efficiency of friendly bacteria in the biofertilizer.
- **Economics of production:** For the production of quality product, use of high-tech instruments and equipments is discernible. However, the low pricing structure and low off take of biofertilizers do not permit the use of sophisticated facilities in a factory. For the same reason, it is also difficult to ensure the production of truly contamination free product. Production of biofertilisers, by and large, is confined to governmental agencies and on a small scale. The problem gets confounded with the small and extremely seasonal demand for these products.

7.2.4.4.2 Marketing constraints

- *Short shelf life:* The shelf life of biofertilizers produced by using the common carries like peat or lignite is often less than six months. The best results are rather possible if the material is used in 3–4 months time, provided the material is not subjected to very high temperature (about 45 °C) during movement and transportation to sale centers or their storage.
- Demand is limited because of inadequate awareness amongst the extension workers, dealers and farmers regarding cost effectiveness and usefulness of biofertilizers.
- Lack of interest on the part of dealers because of slow disposal rate of the product, nominal profit margin and extremely low turnover.
- Lack of adequate promotional and publicity support because of being a low turnover product with limited demand and profitability.
- Unavailability of proper transportation and storage facilities.

7.2.4.5 Precautions to use bio-fertilizers

- Keep the packet (s) of the culture at a cool and shady place, till their use.
- Use the material before the date of expiry mentioned on the packet.
- Use a biofertiliser only for the crop specified on the packet especially in the case of *Rhizobium* culture.
- Open the packet containing the culture just before use and inoculate only that much seed which could be sown immediately.

- Do not put culture in warm or hot water which could destroy the living bacteria contained on the biofertilizer.
- For seed treatment, if besides a bioinoculant, the seed is also to be treated with a fungicide or insecticide, treat the seed first with a fungicide followed by the insecticide and finally with the biofertilizer (bacterial inoculant).
- If the seed is also required to be treated with some mercury-based chemical, the dose of the bioinoculant should be doubled.
- While using a bioinoculant in strongly acidic or saline-alkali soils, it would be desirable to use some soil amendment along with, such as gypsum/phosphogypsum for saline alkali soils and lime or rock phosphate, etc. for strongly soil to maintain soil pH near neutrality (6.5 to 7.5). Else the efficiency of the biofertiliser may be adversely affected.
- Adequate nutrition, for instance, with phosphorus, calcium, etc. for the normal growth and activity of *Rhizobium* is a must and should be restored, Likewise, to meet the requirement of Ca, choice should be in favor of fertilizers like CAN or single superphosphate.

7.2.5 Legumes in Cropping Systems

7.2.5.1 Dual purpose legumes (DPLs)

Legumes have a long standing history of being soil fertility restorer due to their ability to obtain N from the atmosphere in symbiosis with rhizobia. Legumes can form an important component of INM when grown for fodder or grain in a cropping system. A number of leguminous crops have been evaluated for the contribution which they make in meeting the N requirements of the succeeding crop and it has been found that on an average as much as 50–60 kg N/ha may be made available. The carry over of N for succeeding cereal may be 60–120 kg in berseem, 75 kg in Indian clover, 35–60 kg in fodder cowpea, 55 kg in black gram, 60 kg in groundnut, 68 kg in gram and 50 kg in lathyrus. Grain yield of succeeding crop increases markedly when legumes precede them compared with that when cereals proceeded. Pulse crops can be made more effective by inoculation with proper species of *Rhizobia*. In a country where the average consumption of fertilizer is very low, the residual fertility build up due to legumes is obviously a major contribution which must be fully exploited. Pulses have to be used as a source of renewable supply of N.

7.2.5.2 Legume intercropping (LI)

In situations where limitation of time may not permit, inclusion of legumes in rotations, inter-cropping of legumes may be practical solution. Sugarcane is generally planted in rows 60 to 90 cm apart. It normally takes 30–45 days for germination. The initial crop growth up to 100 days after planting is slow. As a result, the lateral spread of crop foliage is not much. Secondly, green manuring prior to sugarcane is considered a wasteful practice because of population pressure on land and availability of fertilizers at cheaper rates. It is not preferred to miss a kharif crop for green manuring. All these factors attracted the attention of scientists and farmers to discover the growing of legumes in inter-row spaces as inter crop or companion crop for grain or green manuring or fodder. Companion cropping of short duration legume crop with autumn planted can is considered as a very useful innovation in economising the resources. The erect growing, short duration and dwarf varieties of legume crop are more suitable for intercropping. Among the summer legumes, moong, urd, and cowpea and popular intercrops with spring planted sugarcane. The green manuring of intercropped legumes has been reported to increase the cane yield. The legumes, even as intercrops have shown potential for N fertilizer economy in terms of increase in the yield of sugarcane. Intercropping of legume increases N use efficiency in sugarcane by checking the losses of NO_3-N and by increasing the N uptake. The utilization efficiency of applied P is also increased by intercropping of legumes.

The success of intercropping of legumes with main crop depends upon the non-competitiveness of the intercrops and if competitive, the extent to which it may compensate the yield loss in main crop, if any, by its produce. The role of legumes has to be assessed in two ways: 1. The accretion of N, 2. Competitive loss to main crop. It may over shadow main crop at a time when the main crop is tilering and may bring initial depression in shoot population. N economy through intercropped legumes is yet to be correctly assessed in different intercropping systems. Several leguminous crops due to their small stature and short duration are ideally suited for introducing as intercrops. These intercrops may be grown for grain, fodder or as green manure. The intercrops, besides increasing the total productivity of the system, also play an important role in economising the resource use especially N. It has been estimated that with inclusion of legume in intercropping system, the extent of N addition would be of the order of 0.746 million tonnes.

7.2.5.3 Legume in rotations

Leguminous crops are known to fix nitrogen, improve soil productivity and benefit the succeeding crop yield compared to non-legume crops in rotation, thus affecting considerable economy in plant nutrients. Studies at Pantnagar showed that wheat after inoculated soybean produced 22.4 q/ha yield compared to 13.5 q/ha after uninoculated soybean. It indicates that inoculation of legumes with suitable bacteria is desirable.

7.3 CONSTRAINTS OF INM

A part from several benefits and importance of INM in sustaining all over agricultural productivity and soil health nation wide, there are some constraints/ limitations to use of INM practices uniquely are given as:

- **Lack of organic materials:** Unavailability of organic materials especially animal manure and crop residues is a primary constraint in many areas.

- **Single multiple enterprises:** Farmers who have less number of cattle may have to depend solely or mainly on fertilizers, whereas farmers who practice several occupations like cash and field crops, together with dairy or livestock, poultry, fisheries enterprise, etc., these enterprises provide opportunities to use/recycle the wastes, manures preferentially and profitably without depending on costly purchased inputs.
- **Peri-urban/rural differences:** Developed markets encourage farmers to use fertilizers and produce more under intensive system of cropping. Even small farmers use more fertilizers inputs. However, in peri-urban areas, there is also possibility for use of agro-industrial or urban municipal wastes along with fertilizers to augment soil fertility. Farmers in remote areas with poor infrastructure and without access to market and also awareness of the benefits of fertilizers may use locally available organic sources.
- **Land tenancy:** The farmers who take land on tenure basis try to harvest high yields using mineral fertilizers and irrigation to ensure a rapid returns to cover the cost of renting the land and may ignore the use of organic manures especially in cereal crop production.
- **High cost of organic manures:** Cost of organic manures especially the animal manures is high in peri-urban areas where these manures are preferentially used in ornamental gardens, lawns and also home gardens in raising vegetable crops.
- **Transport:** Because the organic manures are bulky, it is not convenient to transport and apply them in all crops in all the seasons. So it is applied conveniently in sufficiently good amount in remunerative crops at 4–5 years interval especially in kharif crops.
- **Competitive use of organic resources:** A very important example of competitive use is the use of cow-dung as fuel because of the shortage of fuel wood. Similarly, crop straws or stalks are used as fuel wood especially of the castor, redgram, cotton stalks. Crop residues are also very valuable animal feed. Sometimes poultry manure/droppings are used mixed with other additives and used as fish or cattle feed.
- **Pests, diseases and weeds:** Some believe that the organic manures may carry pests, pathogens and weed seeds and propagate them in the current or following crops.

7.4 DEVELOPMENTAL ISSUES

- Soil test laboratories should be strengthened and upgraded for soil and plant analysis for both macro and micronutrients.
- Greater awareness needs to be created among the farmers about the importance of soil health for monitoring soil nutrient status on regular basis through soil and plant analysis.
- For the promotion of biofertilizers, Government agencies, NGOs as well as corporate sectors should be encouraged to establish biofertilizers production unit

at least at district level. Similarly, such agencies should be provided proper support for the production of compost from city garbage.

- Ensuring availability of good quality fertilizers for micronutrient deficient areas especially zinc, sulphur, etc.
- Fertilizer industries should also be advised to supply biofertilizers along with the chemical fertilizers particularly with respect to nitrogen.
- Greater awareness needs to be created among the farmers on farm resource generation and its proper recycling to serve various rural needs, such as fuel, fodder, manure, etc. through promotion of biogas plant among the farm families. Community approach may also be encouraged for the production of compost and green manure.
- Advantage of introduction of legumes in the cropping systems should be promoted.
- Special efforts to produce the seeds of efficient green manure crops are needed.

7.5 RESEARCHABLE ISSUES

- "Resource management" i.e. strategic use of various nutrient resources is an area of challenge to soil scientists as well as agronomists and considerable part of our future research portfolio must be devoted to this issue.
- Optimization of various resources input combinations specific to particular soil and cropping system better utilization of input resources.
- Improving interactive effects of various components of INM system on various beneficial soil chemicals, physical and biological processes and nutrient use efficiency.
- Highly reliable diagnostic criteria of soil fertility and soil test techniques need to be developed and once, these are in use need to be refined so that their predictability could be pushed up to match the adoption of INM technologies.
- Need to develop model approach for computing INM for major cropping system in each of the agro-sub-ecological zones.
- Energy turnover on the use of INM technology in different production systems.
- Enhancement of shelf life of biofertilizers, development of new strains through genetic engineering and development of easy techniques for viability test of biofertilizers.

7.6 FUTURE PLAN OF WORK

- Considering INM system as a mission for future agriculture, we stress that following few aspects need to be kept in mind to our vision of the future.
- There are many people talking about low external input sustainable agriculture (LEISA) and organic farming in the wake of sustainability debate. This, however, will not serve the future need, and INM technologies must demonstrate that a high input (inorganic and organic) and high yield agriculture can also be sustained.

- Since "production sustainability" is a dynamic phenomenon, all the INM technologies aimed at sustaining productivity of our production system must have a time dimension to indicate the period up to which such technology will remain effective to maintain sustainability.
- INM system should not be used as complementing to synthetic fertilizers; rather it should be developed as an effective nutrient supply and management system for making best use of synthetic fertilizer and other renewable resources of nutrients.
- The all India foodgram production trends indicate the areas that are surplus and deficit. Movement of food grain from the surplus regions to deficit regions offsets the regional nutrient imbalances, thereby turning the intensively cultivated surplus districts regions into areas deficit in nutrient and nutrient imbalances. This trend has to be reversed through INM system.
- Environmental concerns should be given sufficient prominence while developing INM technologies. Ecological sustainability must be the major emphasis so that there is no adverse effect on the other life support systems on long run.
- A computer aided data base needs to be created for all nutrient sources (mineral and organic) including their amount, composition, processing techniques, their economic value and their availability (on-farm-off-farm purchased) to help in planning strategies and managing nutrient efficiently.

7.7 INM OPTIONS FOR IMPORTANT CROPS

Researches on INM in important crops and cropping system in a variety of soils have been conducted throughout the country for proper utilization of plant nutrient sources *vis-à-vis* sustaining crop productivity and soil health are described here serially along with suitable examples for better soil and crop management (Table 7.1).

7.7.1 Rice

It is the most important staple food crop of the country, particularly in humid areas of Assam, Manipur, West Bengal, Orissa, and Eastern UP and in Southern states. It may be grown in a variety of soils like clay loam, red-laterite loamy, Hill Mountain, black soils, etc. The varieties of rice grown in tropics known as *Indica* and in temperate known as *Japonica*. Yield maximization experiments to identify the process and monitor the mechanisms responsible for current stagnation in yield levels clearly indicated that by resorting the conjunctive use of organo-inorganic fertilizers (recommended fertilizer dose + FYM + $ZnSO_4$ + biofertilizers) with optimum plant population per unit area, it is possible to maximize the grain yield of irrigated rice. It has been found that rice yield in newly cultivated areas is much higher than the traditional rice growing area, due to the use of high yielding varieties, better soil and water management practices coupled with integrated nutrient management schedules.

7.7.2 Wheat

Among the cereals, wheat is the most widely cultivated crop in the World and in India. At present India produces more than 72 million tonnes of wheat, which is 11 times higher relative to the same during 1950–51. At present wheat production in India is more than that of United States. The contribution of our country to global wheat production has increased from 10 per cent in nineties to 12 per cent at present. In India, it is grown in zones like North Eastern plains, North Western plains, and Central zone, Peninsular zone, and Northern hilly areas. It was observed that 9.2% of increased grain yield was due to the use of biofertilizers with an average extra yield of 330 kg per ha. As so far integrated nutrient management options are concern in wheat crop, it is recorded at various places that the green manuring, use of caster cake, use of *Azotobactor*, incorporation of FYM, vermicomposting and spent mushroom compost increased wheat grain yield and post harvest available plant nutrients. There frequent uses improving physical and biological properties of wheat growing soils.

7.7.3 Sorghum

It is known for its drought tolerance and is the most popular food/fodder crop in dryland areas of Central and North Indian Zones. Sorghum is grown about 99.8 lakh ha area. The major jowar growing states are MP, AP, Karnataka, Maharashtra, Gujarat, and Rajasthan. Although jowar is a kharif season crop, however, in some Central and Southern part of the country, it is grown as a major *rabi* crop. It can be grown on a variety of soil types, but the clayey loam soil rich in humus is found to be the most ideal soil. It may tolerate mild acidity to mild salinity under pH 5.5 to 8.0. The INM options in sorghum are given in Table 7.1 revealed that the balanced and conjoint use of fertilizers, organics and biofertilizers gave highest fodder and sorghum grain yield at Udaipur, Hyderabad, Paiyur, Faizabad and Akola.

7.7.4 Maize

It is an important staple food crop in India which is grown on plains, hills or mountainous areas having moderate rainfall. In India the major area of the crop is confined to Gujarat, Rajasthan, Punjab, Haryana, MP, UP, AP, HP, Jammu and Kashmir and Bihar. Deep, fertile, rich in organic matter and well drained soils are most preferred for the maize, however may be grown on a variety of soil types. The soil should be medium textured with good water holding capacity. The exploitation of soil as a natural resource has been tremendously increased in the modern agriculture. It is an urgent need to restore the inherent capacity of a soil to supply plant nutrients besides high yield of maize. Integrated nutrient management options for maize are given in Table 7.1 suggested that the use of organics may substituted 25 to 50 per cent of the total nutrient needs of maize crop at Madurai, Coimbatore, Palampur and Meerut.

7.7.5 Pearl Millet (Bajra)

Bajra is grown for grain as well as forage purposes. It is a popular crop grown in areas with less moisture conditions. It is a seasonal kharif crop, but in Central and Southern India and Indogangetic areas, the crop is grown during summer. It is grown on variety of soils such as sandy loams of UP and Punjab and light soils of Rajasthan and North Gujarat, heavy clays of AP, Tamil Nadu and very light soils of Maharashtra and shallow black, red and light soils of Deccan and Southern India. Green manuring along with recommended fertilizer doses found beneficial for improving physico-chemical properties of soil besides yield increment.

7.7.6 Soyabean

Soyabean is grown on a variety of soil types but fertile well drained soils are better. It is grown as kharif crop besides it grows partly in spring season. It prefers neutral soil having a pH between 6.5 and 7.5. It can tolerate mild salinity, but acidity is deadly harmful to the crop. It can be grown on light to black cotton soils of MP. Seed treatment of soybean with *Rhizobium* and PSB found beneficial in increasing yield at Rewa in vertisols (Table 7.1). The integrated use of nutrient sources may save up to 25 per cent of fertilizers.

7.7.7 Groundnut

It is the most important oilseed crop grown in India occupying an area of about 7.572 lakh ha. It is grown mainly in Gujarat, Karnataka, Andhra Pradesh and Tamil Nadu states. The crop requires sandy to loamy soils having drainage facilities and uniform topography. Groundnut also has the commercial value for supplying the raw materials for industrial uses. Integrated nutrient management practices for groundnut crop observed at Rajkot, Mizorum, Secundrabad, Kanpur, Uridhachalam, Sangli, Anantpur and Praphani are given in Table 7.1 for yield and soil health improvement.

7.7.8 Sunflower

Sunflower is an oilseed crop and largely cultivated throughout India. It needs warm humid climate during vegetative growth, bright sunny days during flowering and maturity, but prefers light textured well drained soils. The nutrient requirements of oilseeds like sunflower in general, is high for all the nutrients which need to be supplied in adequate and balanced quantities for high yield. Use of biofertilizer like *Azotobacter, Azospirillum* and phosphate solubilizing microbes can also play an important role in achieving economy in chemical fertilizer use.

7.7.9 Cotton

It is one of the most important fibres yielding crops in India. The cotton contribution to our country's economy is of high significance. It is grown in three agro-ecological zones, viz. Northern (Punjab, Haryana and Rajasthan), Central (Gujarat, Maharashtra and MP) and Southern (AP, Tamil Nadu and Karnataka). All have different species compositions and their zone specific problems. Integrated nutrient management aimed at sustainable crop production by combined use of inorganic fertilizers, organic manure and biofertilizers to promote the inherent nutrient supplying capacity of the soil are need of the time.

7.7.10 Mustard

It is the largest producing country of mustard in the world occupying an area of about 6.5 lakh ha. The major mustard producing states are Rajasthan, UP, Punjab, Haryana, Assam, Bihar, MP, West Bengal and Odisha. It is a dry season crop which requires cool and dry weather with bright sunshine. It is grown under rainfed as well as irrigated conditions. Research on INM indicates that the 25 to 50 per cent of the NPKS needs can be substituted by combination of organic, inorganic and biofertilizers. Besides INM, it is necessary to apply 20–40 kg/ha sulphur and 25 kg zinc sulphate in the sulphur and zinc deficient soils.

7.7.11 Sugarcane

It is the main source of sugar in India. It is commercially grown in Maharashtra, Karnataka, Tamil Nadu, Punjab, Haryana, UP, Bihar, Assam, West Bengal, Orrisa and AP. It requires warm and humid climate for its growth, however, the crop needs high temperature, bright sunlight and high humidity for growth but cool, dry bright sunshine and frostless night during maturity. The crop needs heavy soil, rich in organic matter and plant nutrients, but should be free from salt ideally green manuring and use of sugarcane trash in sugarcane crop helps in increasing urea utilization when combinely used with fertilizers. INM options are given in the Table 7.1 for further reading.

7.7.12 Pulses

These are important for protein in our daily diet. The protein from pulses is easily digestible and relatively cheaper and has higher biological value besides, pulse crop possesses ability to support soil bacteria (*Rhizobium*) in their root nodules which fixes atmospheric nitrogen and enrich soil fertility. Most pulse growing states are Karnataka, AP, Maharashtra, Haryana, Punjab, MP, UP, Uttranchal, Chattisgarh and Rajasthan. Black gram is grown as a spring and summer crop. Green gram is grown as rainfed, during kharif and summer. It grows well on alluvium, red and black cotton soils. Bengal gram is grown in irrigated *rabi* besides in dryland areas. It is also grown as mixed with cereal or millet, but under assured irrigation it is grown as a pure crop. Besides various integrated nutrient management options as suggested in Table 7.1 at various places, it is necessary to use micronutrients like Zn, B, Mo and Fe helps in improving productivity in the following manner.

- Foliar spraying of 0.5 kg $ZnSO_4$/ha with 0.25 limes is effective in correcting Zn deficiency
- Mo deficiency can be corrected by applying 1 kg sodium molybdate/ha

- Soil application of $ZnSO_4$ @ 25 kg/ha to one crop on Zn deficient soils is helpful to both the crops of a pulse based cropping system
- Foliar spray of B @ 1.0–1.5 kg B/ha or soil application of 4 kg borax/ha enhances pod yield on B deficient soils
- Spray of 2% $FeSO_4$ to recoup from Fe deficiency

7.7.13 Vegetables

These are commercially grown in almost all parts of the country and linked by both rich and poor. Presently vegetable cultivation occupies 6.09 million hectares area with an annual production of 84.8 million tonnes. Potato, tomato, brinjal, cauliflower and cabbage are most importantly preferred by public besides okra, onions and pea. The leading states are UP, Bihar, Odisha in North India and Maharashtra, Tamil Nadu and Karnataka in South India. Farmers tend to over use of fertilizers in vegetables. This practice damages their quality. The nutrient supply of each crop is to be balanced by the use of organic manures and fertilizers. For sustainable vegetable production, there must be an optimum nutrient balance. Continuous use of high nitrogenous fertilizers alone cause sharp reduction in soil organic matter. Decline in soil organic matter may be reestablished by the balanced use of NPKS in conjunction with annual application of 10–15 t/ha FYM. INM options in various vegetables are given in Table 7.2 for increasing yield and maintaining soil health. The general INM recommendations in vegetable are follows:

- Addition of organic manure as well as decomposed cowdung, poultry litter and chemical fertilizer recommended.
- Liming at 2 ton/ha every 1 to 2 years is recommended. Lime should be added 2–3 weeks before the addition of chemical fertilizer in acidic soils.
- Crop rotation should be practiced.
- All N fertilizers should be covered with the layers of soil to prevent volatilization losses.

7.7.14 Spices

India is the major producer and exporter of pepper, ginger, turmeric and seed spices. Integrated nutrient management is one of the important factors besides good seed, water, pesticides, which contributes in higher yield of cumin crops, coriander, ginger, and fennel and turmeric crop. Use of organic manures in spice cultivation is must for maintaining quality as well as quantity of spices. Biofertilizer uses have also increased in spices in recent years and fruitful results were observed in several states. A few examples of INM are given in Table 7.2 advised to use all plant nutrient sources in combination gave better yield of spices. Use of organic manure, mulching, crop residues, green manuring and composting.

7.7.15 Fruits

Integrated nutrient management practices were observed in pomegranate, nagpur mandarian, apple and walnut at Bikaner, Nagpur and Shalimar suggested use of inorganic fertilizers alongwith vermicompost, neem cake, dalweed mulch and biofertilizers like *Azospirillum* and *Pseudomonas* improved fruit yield as well as maintained soil fertility.

7.7.16 Ornamental Plants

Flowers are very important part of a human life from worship to aesthetic value. Flowers are grown in almost every states of our country according to climate and needs. Use of integrated nutrient management practices in gladiolus, rose, jasmine and marigold at Allahabad, Shalimar and Bangalore increased the flower yield. It suggested that the INM is as equally important to flower plants as in other cereals, pulses and vegetables.

7.7.17 Miscellaneous crops

The INM practices were also observed in aswangandha, vetiver, foxtail millet, sesame, linseed and potato at various places given in Table 7.2.

7.8 THE INM OPTIONS FOR IMPORTANT CROPPING SYSTEMS (TABLE 7.3)

7.8.1 Rice-Wheat

Rice and wheat grown sequentially in an annual rotation constitute a rice-wheat cropping system. In an annual cycle suitable thermal conditions for both rice and wheat exist in warm-temperate and subtropical areas and high altitudes in the tropics. Total contribution of rice-wheat cropping system in total cereal production is 85 per cent. More than 10 M ha area is occupied by system in India. It is a dominant cropping system in the Indogangetic and non-Indogangetic plains in India. Major growing states are Punjab, Haryana, UP, Uttranchal, Bihar, West Bengal and Himachal Pradesh. Rice is commonly transplanted into puddled soils and prefers continued submergence; wheat is grown in upland well-drained soils having good tilth. The soil test based recommendations of nutrients coupled with organics and biofertilizers proved beneficial for yield as well as soil health. Dhaencha *(Sesbania aculeate)* and *Sesbania rostrate* are good sources of green manures that can be grown in rice-wheat system. GM crops are planted during leg period of 60–75 days between harvest of wheat crop and transplanting of rice after 10 July.

7.8.2 Rice-Rice

When rice is grown after rice annually constitute a rice-rice cropping system which is followed in many states like Assam, West Bengal, etc. The rice straw recycling in such cropping was observed at several places. The rice straw left behind after the previous crop is added to the next crop at recommended rates. A 12–15 per cent increase in yield was recorded over control. About 25 to 50% NPK can be substituted by incorporation of *Sesbania* green leaves; organic manures act as slow releasing fertilizers and prevent the nitrogen losses through leaching and volatilization when used in conjunction with mineral fertilizers.

7.8.3 & 4 Maize-Wheat and Soyabean-Wheat

Integrated use of fertilizers along with green manuring or lantana mulch helps in improving the yield of these cropping systems besides improving soil health. For optimum benefit of INM practices, it is advisable to use micronutrients fertilizers like $ZnSO_4$ @ 25 kg/ha to any component crop in soybean-wheat system in alternate years on Zn deficient soils.

7.8.5 Miscellaneous Cropping Systems

INM options have been recorded in rice-mustard, rice-potato, sugarcane-potato, pearlmillet-wheat, sorghum-gram, groundnut-wheat, sugarcane-wheat, rice-gram, sorghum-chickpea, soybean-winter maize, sugarcane-wheat, sugarcane-lentil, and groundnut-pearlmillet and summer mungbean-winter wheat at various locations of the country focused. A single source of plant nutrients neither fulfilled the all needs of a crop nor provides balance nutrition; hence all sources should be used judiciously accordingly to the need, availability and climatic condition in an integrated manner.

7.9 CONCLUSIONS

INM not only improves soil physical, biological and chemical properties, but increased yield, post harvest availability of nutrients, carbon sequestration and B—C ratio.

Table 7.1 Integrated nutrient management (INM) options for different crops

Regions/Location, Soil type/Soil orderEcosystem/Agro-climatic regions	**INM Options**
Rice (*Oryza sativa* Linn.)	
Rajendranagar (Andhra Pradesh) alfisols, semi-arid	Substitution of recommended nitrogen either 50 or 25 per cent by use of green manure *Glyricidia* gave an additional yield of rice @ 2.71 q/ha and was at par with 100 per cent recommended dose of nitrogen fertilizer (27, 33).
Jabalpur (Madhya Pradesh), clay soil, sub-humid	Incorporation of 5 t/ha of farm yard manure (FYM) and 40 kg N in three splits of 15+15+10 kg and 15 kg P_2O_5/ha help in increasing yield and sustaining soil health (33).
Pusa, Samastipur (Bihar),silty clay loam, sub-humid	Application of 75:45:30 (N:P:K kg/ha) + *prickly sesban* @ 14 t/ha proved effective in increasing rice yield (26).
Kalyani (West Bengal), alluvial, sub-humid	Use of nitrogen 60 kg, phosphorus 30 kg and potassium 30 kg per hectare along with 10 t FYM/ha increased yield and soil health (26).

Contd.

Table 7.1 Integrated nutrient management (INM) options for different crops *(Contd.)*

Regions/Location , Soil type/Soil orderEcosystem/Agro-climatic regions	**INM Options**
Rice (*Oryza sativa* Linn.)	
Imphal (Manipur), submerged	Use of *Azotobacter* at root dip, tillering and booting stage with nitrogen application by urea found beneficial in submerged conditions (35).
Madurai (Tamil Nadu), semi-arid	*In situ* incorporation of green manure, Daincha @ 12 t/ha along with 50 per cent recommended nitrogen with inoculation of *Azospirillum* increased rice yield (2).
Chandrashekarpur (Odisha) rainfed lowland	Application of 75 per cent (60:30:30, N:P:K kg/ha) of recommended dose based on fertilizer along with green manure dhaincha increased rainfed lowland rice yield (2).
Annamalinagar (Tamil Nadu), lowland rice, semi-arid	• Lowland rice yield increased by application of 180:57:57 (N:P:K kg/ha) along with pressmud @ 25 t/ha (26). • Seed treatment with KH_2PO_4 = 100 percent N:P:K + use of *Azospirillum* and *Phosphobacteria* gave highest rice yield (5.72 t/ha) at Annamalinagar (35).
Larnoo (Jammu & Kashmir) alfisol, semi-arid	Incorporation of 20 t FYM/ha along with N:P:K (80:60:40) kg/ha proved beneficial in sustaining crop yield and soil health (44).
Allahabad (UttarPradesh)	Use of flyash @ 40 t/ha along with recommended dose of fertilizer and inoculation with *Azospirillum* proved effective (44).
Ranchi (Jharkhand) Oxisols	Application of 40:30:20 NPK kg/ha in upland rice proved effective in increasing yield as well as nutrient and water use efficiencies (35).
Varanasi (Uttar Pradesh), entisols	Application of 40:30:20 kg NPK/ha increased yield of upland rice (12).
Coimbatore (Tamil Nadu), sandy clay loam, semi-arid	Incorporation of green manure (GM) *Sesbania aculeata* @ 6.25 t/ha or FYM @ 12.5 t/ha along with *Azospirillum* 2 kg/ha and nitrogen @ 150 kg/ha with zinc sulphate @ 25 kg/ha help in sustaining crop yield as well as soil health (46).
Bangalore (Kamataka), sandy loam, semi-arid	Use of STCR dose (75% inorganic + 25% organic (4 t FYM/ha) helps in increase of productivity of rice (35).

Contd.

Table 7.1 Integrated nutrient management (INM) options for different crops *(Contd.)*

Regions/Location , Soil type/Soil orderEcosystem/Agro-climatic regions	INM Options
Rice (*Oryza sativa* Linn.)	
Nellore (Andhra Pradesh), semi-arid	Incorporation of green manures Sunnhemp (5 t/ha) plus soil test based dose of fertilizer increased rice yield (12).
Thandamuthur Udumalpet (Tamil Nadu), sandy loam, arid	Green manure such as *Calotropis* and *Kolingi* with inorganic fertilizers improved productivity of rice and soil (19).
Bhubaneswar (Odisha), sandy loam, sub-humid	Incorporation of 5 t FYM/ha 127:0:41.4 N:P_2O_5:K_2O/ha increased yield of rice and improved soil properties (19).
Durg (Chhattisgarh), vertisols, sub-humid	Combined application of mineral fertilizers + 5 t/ha FYM increased seed yield and soil health (33).
Ropar (Punjab), sandy loam, arid	Application of 124 kg N/ha + 13 t/ha FYM gave highest grain yield (2).
Shalimar (Jammu & Kashmir), silty clay loam, semi-arid	The application of spent mushroom compost (SMC) enhanced the yield of paddy by 19% over control and was found superior to FYM (44).
Wheat (*Triticum aestivum* Linn.)	
Hoshiarpur (Punjab), entisols, arid	Growing of green manure crop of sunhemp in light textured soils, which are generally kept fallow during *kharif*, is an important practice. Incorporation of green manure (GM) crop in the field in the middle of August and raise wheat with the application of 80 kg nitrogen/ha in two equal splits (11).
SK Nagar (Gujarat),loamy sand, arid	Application of 90 kg N/ha through fertilizers + 30 kg N/ha through castor cake gave highest grain yield (26).
Hisar (Haryana),sandy loam, arid	Recommended dose of 120 kg N + 60 kg P_2O_5 + 60 kg K_2O/ha + inoculation of wheat seeds with *Azotobacter* sustained crop productivity (29).
Gurgaon (Haryana),loamy sand, semi-arid	Incorporations of FYM @ 10 t/ha + N120 + P26 + K50 increased grain yield (35).
New Delhi, sandy loam, semi-arid	Incorporation of 90 kg N/ha + FYM @ 10 t/ha increased grain yield (27, 46).

Contd.

Table 7.1 Integrated nutrient management (INM) options for different crops *(Contd.)*

Regions/Location , Soil type/Soil orderEcosystem/Agro-climatic regions	**INM Options**
Wheat (*Triticum aestivum* Linn.)	
Ahmednagar (Maharashtra), sandy loam, semi-arid	20% reduced dose of fertilizer plus FYM @ 2 t/a and *Azotobactor* 20% reduced dose + 5 t FYM/ha + *Azotobacter* (11).
Hisar (Haryana), sandy loam, arid	Use of vermicompost @ 15 t/ha + recommended dose and NPK increased yield of wheat (35).
Hisar (Haryana), sandy loam, arid	Application of 100% NPK (150:60:60) kg/ha + 0.75% spent mushroom compost (SMC) improved yield and soil health (11).
Udaipur (Rajasthan), sandy loam, semi-arid	Cross sowing of wheat + 90 kg N + 40 kg P_2O_5 + 25 kg $ZnSO_4$ /ha increased grain yield of wheat and improved nutrient uptake (31).
Sorghum (*Sorghum vulgare* Pers.)	
Udaipur (Rajasthan), sandy loam, semi-arid	Incorporation of 5 t FYM/ha + N (80%) + 75% (fertilizer) + *Azospirillum* + PSB sustained crop productivity and soil health (31).
Udaipur (Rajasthan), sandy loam, semi-arid	Incorporation of FYM @ 10 t/ha + 75% recommended dose of fertilizers (RDF) + *Azospirillum* + PSB gave highest grain yield of sorghum (31).
Hyderabad (Andhra Pradesh), alfisol, semi-arid	• Application of 2 t/ha *glyricidia* loppings along with 20 kg nitrogen through urea (32). • Application of 4 t/ha FYM + 2 t/ha *glyricidia* loppings. • Application of 4 t/ha FYM + 20 kg N/ha through urea (32). Apart from supplying nutrients, the practices will help in improving soil health by increasing soil organic matter.
Paiyur (Tamil Nadu), loamy sand, semi-arid	Recommended mineral fertilizers + FYM @ 10 t/ha or 75% recommended inorganic fertilizers + biofertilizers increased sorgham grain yield and sustained soil health (35).
Faizabad (UP), inceptisol, semi-arid	Application of 15 kg N through compost along with 20 kg nitrogen through inorganic fertilizer increased yield (46).

Contd.

Table 7.1 Integrated nutrient management (INM) options for different crops *(Contd.)*

Regions/Location , Soil type/Soil orderEcosystem/Agro-climatic regions	**INM Options**
Wheat *(Triticum aestivum* Linn.)	
Akola (Maharashtra),vertisols, semi-arid	Incorporation of 9.45 t/ha of *leucaena loppings* during fallow provides 25 kg N/ha to rabi sorghum (26).
Akola (Maharashtra),vertisols, semi-arid	Application of 50:50:50 kg NPK/ha through pressmud cake and 50 kg N through urea as top dressing increased sorghum yield (26).
Maize *(Zea mays* Linn.)	
Satpura (Madhya Pradesh), lighter soil of satpura plateau	Combined use of 5 t/ha vermicompost + NPK (RDF) recorded 24.35 q/ha grain yield (26).
Chindwara (MadhyaPradesh), sub-humid	Application of 90:40:30, $N:P_2O_5:K_2O$ kg/ha + FYM 10 t/ha increased grain yield and sustained crop productivity (35).
Madurai, (Tamil Nadu), sandy clay loam, semi-arid	Application of 75% RDF through fertilizer + 5 t FYM/ha increased higher grain yield of maize (32).
Larnoo (J & K), alfisol	Use of 80 kg N + FYM @ 10 t/ha recorded highest grain yield (44).
Rakh Dhiansar, entisol	Application of 60-40-20 kg NPK/ha + 20 kg/ha zinc sulphate improved yield in rainfed conditions (26).
Coimbatore (Tamil Nadu), medium black	Application of 75% N through vermicompost along with 25% N through neem cake improved yield and soil health (26).
Palampur (Himachal Pradesh), clay loam	Application of FYM can substitute 50% of recommended nitrogen helps in economizing 50% of recommended NPK in maize (24).
Modipuram (UP), sandy loam, semi-arid	Application of 50% recommended dose through fertilizer and remaining 50% N as FYM helps in sustaining crop productivity and soil health (31).
Pearl millet *(Pennisetum typhodeum* Linn.)	
Hisar (Haryana), sandy loam, arid	FYM @ 2.5 tonnes/ha + 20 kg N/ha + seed treatment with biofertilizers help in increasing yield and sustained soil health (33).
Durgapura (Rajasthan), sandy loam, semi-arid	Application of 5 t/ha vermicompost or 100% RDF proved beneficial in rainfed pearl millet crop (35).

Contd.

Table 7.1 Integrated nutrient management (INM) options for different crops *(Contd.)*

Regions/Location , Soil type/Soil orderEcosystem/Agro-climatic regions	INM Options
Pearl millet ***(Pennisetum typhodeum*** **Linn.)**	
Jamnagar (Gujarat), medium black	Incorporation of 5 t FYM/ha + 40 kg N/ha increased grain yield of bajra (33).
Hisar (Haryana), sandy loam	Application of 120 kg N + 60 kg P_2O_5 along with vermicompost @ 5 t/ha recorded highest seed yield (2270 kg/ha) of rainfed pearl millet (33).
Rahuri (Maharashtra), medium black, semi-arid	Incorporation of green leaf manuring one month before sowing (5000 kg/ha subabul or *glyricidia)* or intercropping with cowpea and sunnhemp improved soil physico-chemical properties of soil (33).
Agra (Uttar Pradesh) entisol	Application of 40 kg N/ha through fertilizer help in increasing yield (33).
Hisar (Haryana), sandy loam, semi-arid	Application of 94:60:0:25 kg $N:P_2O_5:K_2O:ZnSO_4$ along with incorporation of 1.5 t FYM/ha in kharif pearl millet recorded highest grain yield (35).
Coimbatore (Maharashtra), sandy loam, semi-arid	Use of 75 kg N/ha as urea + FYM @ 2.5 t/ha increased grain yield (35).
Coimbatore (Tamil Nadu), clay loam, semi-arid	Use of *Azospirillum* + 100% NP + *phosphor-acterium* contained crop and soil productivity (35).
Leh (J & K), sandy loam, sub-humid	Application of 60 kg N/ha along with FYM 10 t/ha improved grain yield and soil health (44).
Soybean (*Glycine max* Merr.)	
Coimbatore (Tamil Nadu), sandy loam, semi-arid	Application of 75% recommended nitrogen along with biofertilizers found beneficial for better yield and soil health (2).
Sagar (UP), Medium black, semi-arid	Application of 100% N:P:K (20:80:20 kg/ha) recorded highest seed yield/plant (2).
Indore (MP), vertisolsemi-arid	Incorporation of FYM @ 6 t/ha along with 20 kg nitrogen and 30 kg phosphorus/ha increased seed yield and soil health in indore region (31).
Rewa (MP), vertisol, semi-arid	Seed treatment of soybean with *Rhizobium* bacteria and phosphate solubilizing bacteria (PSB) helps in increasing yield of soybean (2).

Contd.

Table 7.1 Integrated nutrient management (INM) options for different crops *(Contd.)*

Regions/Location , Soil type/Soil orderEcosystem/Agro-climatic regions	INM Options
Soybean (*Glycine max* Merr.)	
Bhopal (MP), black soil,sub-humid	Balanced fertilization based on soil test along with incorporation of FYM @ 4 t/ha increased yield of soybean and sustained soil health (33).
Udhamsingh nagar (Uttaranchal) clay loam, sub-humid	Application of 25 kg N + 5 tonnes FYM/ha or 25 kg N + 1 tonnes neem cake/ha gave highest yield (35).
Bapatla (AP), sandy loam, sub humid	Use of biogas slurry 15 t/ha + 50 kg N/ha as urea help in sustaining crop productivity (2).
Udaipur (Rajasthan), sandy loamsemi-arid	Seed treatment of soybean with *Rhizobium* culture @ 5 g/kg of seed arid soil inoculation PSB @ 500 g/ha by mixing with 50 kg of well decomposed FYM at the time of sowing increased fertilizer uptake and yield of soybean and sustained soil health (26).
Groundnut (*Arachis hypogea* Linn.)	
Rajkot (Gujarat), vertisols, arid	Application of NPK (6:12:0 kg/ha) along with mulching of sunhemp in between rows plus *Rhizobium* and PSB treatment gave higher pod yield (31).
Mizorum (Meghalaya), hapludult, sub-humid	Application of 100% NPK (full recommended dose) combined with pig manure @ 5 t/ha recorded 20.02 q/ha yield and improved physical properties of groundnut growing soils (2).
Secunderabad (AP), sandy loam, semi-arid	Application of 100% NPK (full recommended dose) combined with inoculation *Rhizobium* @ 2 kg/ha and *phosphobacteria* @ 2 kg/ha as seed treatment recorded higher yield and sustained soil health (33).
Kanpur (UP), sandy loam, semi-arid	NPK (20:30:45 kg/ha) with incorporation of FYM @ 10 t/ha + gypsum application @ 300 kg/ha for groundnut crop proved beneficial (33).
Vridhachalam (Maharashtra), alfisols, semi-arid	Incorporation of FYM @ 10 t/ha combined with biofertilizer (*Rhizobium* + *phospho-bacterium*) helps in increasing crop productivity and soil health (33).

Contd.

Table 7.1 Integrated nutrient management (INM) options for different crops *(Contd.)*

Regions/Location , Soil type/Soil orderEcosystem/Agro-climatic regions	**INM Options**
Groundnut *(Arachis hypogea* **Linn.)**	
Sangli (Maharashtra), black soils semi-arid	Incorporation of FYM @ 2 t/ha + 75% recommended dose of NPK to kharif groundnut increased pod yield (31).
Anantpur (Maharashtra), alfisol	Use of *Rhizobium* @ 50 gram/ha with N, P_2O_5, K_2O (20:40:40) kg/ha increased groundnut yield at Anantpur (26).
Parbhani (Maharashtra), clay, semi-arid	Application of 50% NPK through chemical fertilizers and 50% NPK through inorganic plant nutrient sources (FYM + *Azotobactor* + cow-dung urine slurry + phosphate solubilizing bacteria) sustained soil and groundnut production (26).
Sunflower ***(Halianthus annus*** **Linn.)**	
Hisar (Haryana), sandy loam, semi-arid	Application of 30 kg N/ha through fertilizer along with incorporation of vermicompost @ 7.5 t/ha with *Azotobactor* increased crop productivity and soil health (2).
Dharwad (Karnataka)	Use of 100% recommended dose of fertilizer along with either FYM or vermicompost or poultry manure increased sunflower seed yield (31).
Hisar (Haryana), sandy loam, semi-arid	Application of 30 kg N/ha with incorporation of FYM 10 t/ha increased sunflower yield (35).
Coimbatore (TN), sandy loam, semi-arid	Application of 75% recommended nitrogen through fertilizer along with use of biofertilizers (*Azotobactor* and PSB) helps in increasing yield and soil health (33).
Hyderabad (AP), sandy loam, semi-arid	Use of P-enriched FYM (phosphocompost) + *Azospirillum* + *phosphobacteria* with 0.2% borax helps in increasing yield of sunflower (23).
Shalimar (J & K), silty clay loam	Use of 75% recommended dose of NPK along with 10 t/ha FYM proved beneficial (44).
Hyderabad (AP), alfisols	Application of 5–10 t/ha FYM in conjunction with 50% recommended dose of NPK offers enhanced yield and nutrient use efficiency (23).

Contd.

Table 7.1 Integrated nutrient management (INM) options for different crops *(Contd.)*

Regions/Location , Soil type/Soil orderEcosystem/Agro-climatic regions	INM Options
Cotton ***(Gossypium*** **spp.)**	
Bhawaniputna (Odisha), clay soil, sub-humid	Application of N, P and K @ 100, 50 and 50 kg/ha in conjunction with *Azotobacter* and PSB @ 5 kg/ha each along with incorporation of FYM @ 10 t/ha increased cotton boll yield and sustained soil health (31).
Hisar (Haryana), sandy loam, semi-arid	Application of 50% NPK recommended dose through fertilizers and 50% nitrogen through FYM recorded highest gross return in cotton (35).
Akola (Maharashtra), clay loam, semi-arid	Application of 50 kg N and 25 kg P_2O_5/ha through urea and single superphosphate (SSP) along with FYM @ 10 t/ha helps in increasing yield and soil health (2).
Nagpur (Maharashtra), medium black soil, semi-arid	Use of biofertilizer strains of *Azotobactor*, PSB and VAM based INM packages proved beneficial to crop and soil in cotton black soils (16).
Nagpur (Maharashtra), black soils, semi-arid	Application of 50% NPK of recommended dose along with incorporation of FYM @ 10 t/ha helps in sustaining crop and soil productivity (2).
Parbani (Maharashtra), vertisols, semi-arid	Combined application of 50% recommended dose through chemical fertilizers and 50% recommended dose through FYM and seed treatment with *Azotobactor* increased cotton boll yield and sustained soil health (35).
Mustard ***(Brassica*** **spp.)**	
Udaipur (Rajasthan), sandy loam, semi-arid	Application of 75% NPKS recommended dose based on soil test along with incorporation of FYM @ 10 t/ha or 75% NPKS based on soil test along with the use of biofertilizer *(Azotobactor* + PSB) helps in increasing seed and oil yield of mustard and sustained sole health. Both INM practices proved effective in increasing soil physico-chemical and microbial properties of soil (34, 38 & 39).
Kalyani (W. Bengal), alluvial, sub-humid	Combined application of N:P:K (20:100:100 kg/ha) and FYM @ 10 t/ha increased seed yield of mustard and improved soil physical properties (11).

Contd.

Table 7.1 Integrated nutrient management (INM) options for different crops *(Contd.)*

Regions/Location , Soil type/Soil orderEcosystem/Agro-climatic regions	INM Options
Mustard *(Brassica* spp.*)*	
Jamnagar (Gujarat)medium black	Application of 50% NPK along with seed treatment with biofertilizers (PSM + *Azotobactor)* gave highest yield of mustard (23).
S.K.Nagar (Gujarat), calcareous and saline soil arid	On-farm trials conducted on saline farmers field showed that application of caster cake 1 t/ha increased mustard yield by 14.1 percent over farmer practice (33).
Sugarcane (*Saccharum officinarum* Linn.)	
Gangavati (Karnataka), vertisols, arid	Application of 125% recommended dose of fertilizer along with pressmud @ 3.5 t/ha increased cane yield (47).
Madurai(Tamil Nadu)	Incorporation of composted trash @ 5 t/ha + *Azotobactor* + recommended levels of NPK improved cane yield and soil physical properties of sugarcane growing soils (47).
Shahjahanpur (UP)	Green manuring is an important practice in these areas, and 50 percent of the benefit of green manuring could be obtained in the form of increased sugarcane yield (47).
Modipuram (UP) sandy loam, semi-arid	When organic manures like FYM, crop residues, poultry manure, etc. are used along with fertilizers, the efficiency of urea utilization by sugarcane crop increase and help in better soil health (26).

Table 7.2 INM options for pulse, vegetables, spices, fruits, ornamental plants and miscellaneous crops

Name of crops	INM options
Pulse crops	
Chickpea (*Cicer arietinum* Linn.)	• At Junagadh (Gujarat) application of 20 kg N + 40 kg P_2O_5/ha + *Rhizobium* inoculation helps in increasing pod yield and sustaining soil health in medium black soils (26). • At Sikundrabad (AP) in semi-arid ecosystem, application of 100% NPK + *Rhizobium* 2 kg/ha + phosphobacteria @ 2 kg/ha improved the yield and sustained soil health in sandy loam soils (2).

Contd.

Table 7.2 INM options for pulse, vegetables, spices, fruits, ornamental plants and miscellaneous crops *(Contd.)*

Name of crops	**INM options**
Pulse crops	
Chickpea (*Cicer arietinum* Linn.)	• At Udaipur (Rajasthan) in sandy loam soils, application of N (50%) + P_2O_5 (50%) through inorganic fertilizer along with *Rhizobium* + PSB + VAM sustained crop productivity and soil health (31). • At Bellary in vertisols, application of 15 kg N through green leaf + 10 kg N through inorganic fertilizer increased pod yield of chickpea (31).
Pigeon pea (*Cajanas cajan* Milsp.)	• At Akola (Maharashtra) in vertisols, application of FYM @ 5 t/ha + 40 kg P_2O_5/ha + microbial culture @ 15 kg/ha to pigeon pea (11). • At Sehore (MP) use of 50% N through urea and 50% N through pressmud/cotton cake/castor cake compost + recommended P_2O_5 and K_2O increased pod yield of pigeon pea in vertisols (26).
Green gram (*Vigna radiata* Boxb.)	Application of 75% recommended dose through inorganic fertilizer combination with use of poultry manure (25%) dose gave the highest yield (11.36 g/ha) at Allahabad (U.P.) in sandy loam soils (11).
Peas (*Pisum sativum* Linn.)	At Palampur (HP) in clay loam soil of sub-humid ecosystem, application of N:P:K (20:60:30 kg/ha) along with incorporation of FYM @ 20 t/ha improved pea yield and soil properties (35).
Cow pea (*Vigna sinensis* Savi)	Application of 75% recommended N and biofertilizer increased cowpea yield at Coimbatore (Tamil Nadu) in sandy soil (11).
Black gram (*Vigna mungo var. radiatus* Linn.)	• At Coimbatore (Tamil Nadu) application of 75% P_2O_5 as *Tunisia* rock phosphate + vermicompost @ 2 kg/ha + phosphobacteria @ 2 kg/ha improved crop and soil productivity (2). • At Marutera (AP) in vertisols inoculation of VAM + 50 kg P_2O_5/ha increased pod yield of blackgram (33).

Contd.

Table 7.2 INM options for pulse, vegetables, spices, fruits, ornamental plants and miscellaneous crops *(Contd.)*

Name of crops	INM options
Pulse crops	
Green gram (*Vigna radiata* Roxb.)	• Application of 10 kg N/ha through urea under conventional tillage along with incorporation of 2 t/ha FYM increased mungbean yield at Hyderabad (AP) in alfisols (2). • Incorporation of FYM @ 2 t/ha along with *Glyricidia loppings* @ 1 t/ha under reduce tillage improved crop productivity and soil health (26).
Vegetable crops	
Brinjal (*Solanum melongena)*	Application of 50 kg nitrogen through urea along with 50 kg nitrogen through poultry manure per hectare help in increasing yield of brinjals and sustained soil health (44).
Okra (*Abalmosms caculantus)*	• Application of 20 kg nitrogen by ammonium sulphate with 20 kg nitrogen through poultry manure per hectare increased fruit yield of okra (44). • At Knonkan (Gujarat) incorporation of FYM @ 15 t/ha along with N, P_2O_5 and K_2O (100:50:20) or biofertilizer (Plantrich) @ 375 kg/ha increased fruit yield and help in sustaining soil health (44). • Combination of organic (FYM @ 10 t/ha) and inorganic (50:40:25 NPK kg/ha) found beneficial in increasing yield (1, 20).
Tomato (*Lycopersicon esculentum* Mill)	Incorporation of FYM @ 40 t/ha along with half dose of NPK (75:30:30) kg/ha substituted 50% of recommended dose of fertiliser (19).
Onion (*Allium cepa* L.)	At Wadura of Jammu and Kashmir application of 52.5 kg nitrogen per hectare with the use of *Azotobacter* biofertilizer recorded higher onion yield (44).
Garlic (*Allium sativum* L.)	At Shalimar Srinagar in silty clay loam soil inoculation of seed with *Azotobacter* and phosphobacteria + application of 75 kg nitrogen and 45 kg phosphorus helps in better yield and soil health (44).

Contd.

Table 7.2 INM options for pulse, vegetables, spices, fruits, ornamental plants and miscellaneous crops *(Contd.)*

Name of crops	INM options
Vegetable crops	
Chilli (*Capsicum annum* L.)	At Dapoli (Maharashtra) post or FYM or Glyricidia @ 10 t/ha + 50% recommended dose of fertilizer proved beneficial (35).
Cabbage (*Brassica olerecia* var *capitata*)	• Incorporation of FYM @ 10 t/ha along with fertilizers (N:P_2O_5:K_2O) 150:60:60 kg/ha increased cabbage yield and sustained soil health at Shalimar in silty clay loam soil (44). • At Coimbatore (Tamil Nadu) in sandy clay soil application of 75% recommended nitrogen along with inoculation with biofertilizer proved beneficial (2).
Potato (*Solanum tuberosum*)	• Application of 75% recommended fertilizer along with incorporation of 20 t/ha FYM increased tuber yield of potato in clay soils of Palampur (30). • At Kalyani (West Bengal) in alluvial soil application of N:P:K (60:30:30) kg/ha along with incorporation of FYM @ 10 t/ha recorded higher tuber yield (2). • At Shimla (HP) incorporation of vermicompost @ 5 t/ha found beneficial and recorded (65 q/ha tuber yield) proved the importance of vermicompost in tuber crops (24, 25). • At Shalimar in silty clay loam soil application of 75% recommended dose of NPK along with treatment of tuber with 1% urea and 1% sodium bicarbonate helps in better quality production of potato (44). • At Shalimar in silty clay loam soil incorporation of FYM @ 30 t/ha + fertilizer N:P:K @ 120:60:60 kg/ha increased tuber yield (44).
Broccoli (*Brassica oleracea* L. var. *Halicaplencic*)	Application of FYM + digested sludge (each @ 10 t/ha) and seedling inoculation with VAM only noticed significant improvement in fresh and dry weight of head, head yield at Varanasi (44).

Contd.

Table 7.2 INM options for pulse, vegetables, spices, fruits, ornamental plants and miscellaneous crops *(Contd.)*

Name of crops	INM options
Spices Crops	
Cumin crops (*Cuminum cyminum* L.)	At Ajmer (Rajasthan) use of *Azospirillum* + sheep manure @ 10 t/ha + vermicompost @ 7.5 t/ha increased cumin yield and improved soil physical and microbiological properties (2).
Coriander (*Coriandrum sativum*)	At Kota (Rajasthan) application of 50% nitrogen through chemical fertilizer and 25% nitrogen through compost + 25% nitrogen through *Azotobacter* + use of zinc @ 5 kg/ha (26).
Ginger (*Zingiber officinure* Rosc.)	Use of Azofert @ 20 kg/ha gave highest ginger yield (192 q/ha), and net returns (₹/ha) at Palampur in silty clay loam soils having pH 6.83 (44).
Cumin (*Cuminum cyminum*)	Use of 10–15 t/ha FYM + 30 kg N + 20 kg P/ha improved the yield and quality of cumin (11).
Fennel (*Foeniculum vulgare* Mill.)	Incorporation of 10 t/ha FYM along with 80 kg N + 45 kg P increased yield and soil health (24 & 25).
Turmeric (*Curcuma longa* L.)	20 t FYM/ha + 50 kg P_2O_5 + 50 kg N/ha + 100 kg K_2O/ha improved crop and soil health (2).
Fruit Crops	
Pomegranate (*Punica granatum*)	At Bikaner (Rajasthan) use of inorganic fertilizers along with vermicompost helps in increasing yield as well as soil properties (26).
Nagpur mandarian (*Citrus* spp.)	At Akola (Maharashtra) in medium black soils, application of 800 g N, 300 g P_2O_5 and 600 g K_2O/plant along with neem cake @ 75 kg/tree/year increased fruit yield and sustained soil health.
Banana (*Musa paradisica*)	Any kind of green manure, animal waste (solid/liquid), kitchen waste, crop residues or even high C — N ratio material such as coir dust/saw dust/rice husks can be used as a basal organic material/amendment for banana crop. If material/amendment used have high C — N ratio, it is advisable to use low C — N ratio materials like poultry manure, cattle manure, etc. to achieve quick decomposition *in situ*. It is a shallow rooted crop (45–60 cm). Therefore, application of any kind of easily decomposable materials as a top dressing at 3 months interval is highly beneficial (20).

Contd.

Table 7.2 INM options for pulse, vegetables, spices, fruits, ornamental plants and miscellaneous crops *(Contd.)*

Name of crops	**INM options**
Fruit Crops	
Apple (*Malus domestica* Borkh)	In apple, highest fruit yield was obtained when antitranspirant (salicylic acid) was combined with dal-weed mulch (44).
Walnut (*Juglans regia* L.)	The application of *Azospirillum* + 10 g N plant^{-1} and *Psedomonas* + 5 g phosphorus and leaf area as compared to only N @ 15 g/ plant and only P @ 5g/plant, respectively at Shalimar, J & K (44).
Pineapple	A suitable and freely available organic amendment must be applied as a basal dressing into the planting furrows, e.g. plant residues and animal waste. Foliar application of liquid fertilizer during dry period is more beneficial. Crop residues from previous pineapple crop could be composted and utilized for the next crop (14).
Ornamental plants /place or location	
Gladiolus (*Gladiolus hybrida* L.) Allahabad (UP)	The combined application of organic manures and inorganic fertilizers, i.e. 75% RDF + FYM + VCM + VAM resulted in the higher plant growth, plant height number of leaves/plant, increased number of florets/spike, improved flowering and increased length of spike (2).
Rose (*Rosa* spp.) Karnataka, Jammu & Kashmir Sub humid and semi-arid	A marginal influence in flower yield was recorded with the use of *Azotobactor* as compared to control. Uses of biofertilizer cannot be ignored since much chemical fertilizers were saved besides the yield was at par with control (44).
Jasmine (*Jasminum* spp.) Karnataka, Kashmir, Arid, semi-arid	A use of biofertilizer coupled with organic manure (FYM, crop residues) proved beneficial and increase the flower yield by 26 percent over control (44).
Marigold (*Tagetus* spp.) Kashmir, silty clay loam, semi-arid	*Azotobactor* and PSM biofertilizers were tested in Kashmir to observe the potentialities to bear flowers in marigold and result recorded, indicate the 17 percent increased flower yield over control. Use of biofertilizers sustained flower productivity as well as soil health (44).

Contd.

Table 7.2 INM options for pulse, vegetables, spices, fruits, ornamental plants and miscellaneous crops *(Contd.)*

Name of crops	INM options
Miscellaneous crops	
Aswangandha	In black soils of indore incorporation of FYM @ 2.5 t/ha along with application of 12.5 kg N and 25 kg P_2O_5/ha increased the aswangandha yield (26).
Vetiver (*Veteveria ziaanioides* L.)	In shallow black soils of Malwa (MP) application of N (75) + P_2O_5 (40) kg/ha along with *Rhizobium* @ 2 kg/ha resulted in increased net profit and B:C ratio (26).
Foxtail millet (*Eleusine oracona* Gaertn)	At Bangalore in sandy loam soils, application of N (22.5 kg) + P_2O_5 (11.25 kg) + enriched FYM @ 7.5 t/ha increased yield and foxtail millet (35).
Sesame (*Sesamum indicum* Linn.)	At Jalgaon (Maharashtra) in vertisols, application of 50% recommended N through urea + 50% N through pressmud/cotton cake/ castorcake/compost along with recommended P and K increased crop yield and sustained soil health (35).
Linseed (*Linum usitatissimum* Linn.)	At Kota (Rajasthan) application of N:P:Zn:S (90+30+5+50 kg/ha) along with inoculation of biofertilizer (*Azotobacter* + PSB) and FYM @ 9 t/ ha increased linseed yield and soil health (26).
Potato (*Solanum tuberosum*)	Highes tuber yield in potato (250.57 q/ha) was obtained with 75% NP + seed treatment with urea + sodium bicarbonate + biofertilizers at Shalimar, J & K (44).
Tea (*Camellia* spp.)	The use of organic manures in the form of digested coirpith compost (DCC) and biofertilizers like *Azospirillum brasilense*, Vesicular Arbuscular Mycorrhizae (VAM) and phosphobacteria increased nutrient content and certain enzyme activities in tea hence increased productivity (23).

Table 7.3 INM options for different cropping systems

Place/soil type/soil order/ecosystem	INM options
Rice wheat cropping system	
Ludhiana (Punjab), loamy sand soil, semi-arid	75% NPK alongwith 25% N supply by Jantar green manure to rice crop and 100% NPK for wheat crop sustained crop yield and soil health (24).
Hoshiyarpur and Kapurthala (Punjab) loamy soil and clay loam, sub-humid	Conjoint use of 75% NPK along with 25% green manure or FYM @ 6 t/ha for rice and 75% NPK to wheat found at par with 100% NPK to rice and wheat. It indicates a net saving of 25% NPK by using green manure and FYM in rice-wheat cropping system (24).
Jabalpur (MP), vertisols, sub-humid	Combine application of either 50% recommended dose of NPK or 100 % recommended dose of NPK along with green manure for rainy season rice and 100% recommended dose of NPK along with FYM 6 t/ha for wheat proved beneficial in rice-wheat cropping system (35).
Sambalpur (Odisha), sandy clay loam, sub-humid	Incorporation of green manure and 50% recommended dose of NPK to rice crop and 100% NPK to wheat crop improved soil health and crop yield (35).
Meerut (UP), sandy loam, semi arid	Combined use of fertilizer @ 75% NPK + 25% N as FYM/PSM to rice and 100% NPK to wheat crop (24).
J & K, HP, alluvial, sandy loam	Application of 40 kg N + FYM/ GM @ 15 t/ha + 20 kg $ZnSO_4$ to rice and 120 kg N + 80 kg P_2O_5 (through SSP) + 40 kg K_2O to wheat crop improved productivity of rice-wheat cropping system and sustained soil health (24).
North Bengal, Assam, soils are alluvial, red and brown hill with acidic reaction	Combine use of chemical fertilizer (40 kg N+20 kg P_2O_5 + 40 kg K_2O) + FYM @ 5 t/ha/GM + Azolla @ 10 t/ha + 20 kg $ZnSO_4$ once in 3 years. 5 kg borax to rice and 50 kg N + 20 kg P_2O_5 + FYM @ 5 t/ha to wheat crop found beneficial (12).
Plains of Bengal, red and laterite	Use of 40 kg N + 45 kg P_2O_5 + 30 kg K_2O + FYM/GM @ 10 t/ha + Azolla @ 10 t/ha/ BGA @ 10 kg/ha + 20 kg $ZnSO_4$ to rice crop and 90 kg N + 45 kg P_2O_5 + 45 kg K_2O to wheat crop (12).
Bihar, Eastern UP, alluvial red and laterite	Integrated use of fertilizers (40 kg N + 30 kg P_2O_5 + 20 kg K_2O) along with GM (green gram stover) + 20 kg $ZnSO_4$ in calcareous soil to rice and 90:60:30 ($N+P_2O_5+K$) + FYM @ 10 t/ha to wheat (12).

Contd.

Table 7.3 INM options for different cropping systems *(Contd.)*

Place/soil type/soil order/ecosystem	INM options
Rice wheat cropping system	
Western UP, alluvial	Integration of fertilizers (90 kg N + 30 kg K_2O) + FYM/GM (Sesbania/Leucaena ropping) @ 10 t/ha to rice crop and 90 kg N + 60 kg P_2O_5 + 30 kg K_2O to wheat crop (24).
Lucknow (UP), alluvial	Integrated management of green manuring with *Sesbania rostrata* and application of BGA @ 12.5 kg soil based inoculum and chemical fertilizer has emerged a promising packages for nitrogen management for yield optimization in rice and 100% NPK to wheat crop improved productivity of rice-wheat cropping system and sustained soil health (27).
Hisar (Haryana), sandy loam, arid	Soil test based fertilizers dose + 7.5 t FYM/ha or 15 t FYM/ha to rice-wheat (11).
Rice-rice cropping system	
Assam, Bengal,alluvial, red and brown hill with acidic reaction	Combine use of chemical fertilizer @ 20 kg N + 20 kg P_2O_5 + 15 kg K_2O + FYM/ GM @ 10 t/ha + Azolla @ 10 t/ha + 20 kg $ZnSO_4$ once in 3 years and 60 kg N + 40 kg P_2O_5 + Azolla @ 10 t/ha in rice-rice system increase and productivity and sustained soil health (24).
Plains of Bengal,alluvial and red soil	Application of 60 kg N + 40 kg P_2O_5 + 30 kg K_2O + FYM/GM @ 10 t/ha + 20 kg zinc sulphate and 90 kg N + 80 kg P_2O_5 + 60 kg K_2O + Azolla @ 10 kg/ha (12).
Coastal plains and Ghats and Deccan Plateau, lateritic,alluvial and black	Incorporation of FYM/GM @ 5 t/ha coupled with 75 kg N + 15 kg P_2O_5 and 15 kg K_2O and Azolla @ 10 t/ha + BGA @ 10 t/ha + 90:60:40 NPK improved yield and soil health (12).
Maruteru (AP), clay loam, semi-arid	50% recommended NPK through fertilizers + 50% through seabania green leaf in rainy season and 100% RDF of NPK through fertilisers (12).
Jorhat (Assam) alluvial, sandy loam	INM packages consisting of organic manure @ 1–3 t/ha (dry weight basis) + *Azospirillum* + PSB + P_2O_5 @ 10 kg/ha as rock phosphate and K_2O @ 10–20 kg/ha considered as a suitable alternative for nutrient management in rice-rice system (24).
24 parganas (W Bengal)	Depending upon the organic matter and nitrogen status of soil, *Azolla* and BGA biofertilizers, either alone or in combination can be used to optimise rice yield in both kharif and summer seasons at 20–25% less use of N-fertilizers from that of recommended in alluvial and laterite soils of West Bengal (12).

Contd.

Table 7.3 INM options for different cropping systems *(Contd.)*

Place/soil type/soil order/ecosystem	**INM options**
Rice-rice cropping system	
Puducherry, clay loam	In kharif rice-summer rice combine use of BGA + *Azospirillum* in combination for N nutrition in rice might substitute 20–25% chemicl fertilizers N use from recommended (80–100 kg/ha) without any loss in yield. Yield optimization response of biofertilizers was less in soils showed higher response to high dose of N as chemical fertiliser (12).
Kharagpur (W Bengal), sandy clay loam, sub-humid	75% NPK + 25% FYM to rainy season and 100% NPK to winter season improved soil fertility and plant nutrition (24).
Maize-wheat cropping system	
Jammu and Kashmir, HP and Uttranchal, alluvial, brown hilly and sandy soil	Integrated use of fertilizers (60 kg N + 30 kg P_2O_5 + 20 kg K_2O) along with 10 t/ha FYM + fresh Eupatorium/ antana mulch @ 10 t/ha to maize and 80 kg N + 30 kg P_2O_5 + 15 kg K_2O to wheat crop increased yield and soil health in maize-wheat system (12).
Bihar, Eastern UP, alluvial with neutral alkaline reaction soil	Combine application 60 kg N + 60 kg P_2O_5 + GM + 16 kg borax in calcareous soil to maize and 90 kg N + 60 kg P_2O_5 + 30 kg K_2O with FYM @ 10 t/ha proved effective (12).
Western UP, alluvial	Application of 50 kg N + 20 kg K_2O/ha + FYM @ 10 t/ha to maize and 120 kg N + 60 kg P_2O_5 + 40 kg K_2O to wheat crop sustained soil health and crop yield (24).
Himachal Pradesh, brown hill soil	Application of FYM to meet 50% of recommended N helps in economizing 50% of recommended NPK in maize under maize-wheat system. Maize receiving FYM requires only half of the recommended dose of fertilizers to produce yields equal to full RDF dose applied as fertilizer (12).
Soybean-wheat cropping system	
Rajasthan, MP,alluvial, red and black soils	Application of 10 kg N + 25 kg P_2O_5 (through boronated SSP) + FYM @ 4 t/ha + *Rhizobium* + 25 kg zinc sulphate in alternative year to soybean and 90 kg N + 45 kg P_2O_5 to wheat crop (35).
Maharashtra and Southern MP	In soybean crop 10 kg N + 25 kg P_2O_5 + 4 t/ha FYM and 90 kg N + 45 kg P_2O_5 to wheat crop help in sustaining soil health and crop yield (35).

Contd.

Table 7.3 INM options for different cropping systems *(Contd.)*

Place/soil type/soil order/ecosystem	INM options
Soybean-wheat cropping system	
Indore (MP), vertisols, semi arid	Substitute 50% of fertilizers N through 10 t FYM/ha in soybean-wheat system (24).
Bhopal (MP), vertisols, semi-arid	Application of 8 t/ha FYM to soybean and 60 kg N + 11 kg P/ha to wheat or application of 4 t FYM/ha + 10 kg N + 11 kg P/ha to soybean and 90 kg N + 22 kg P/ha to wheat gave the yield of 2 t/ha soybean and 3.5 t/ha wheat besides sustaining soil health (2).

Table 7.4 INM options for miscellaneous cropping systems

Regions/location soil type/ soil order ecosystem/agro-climatic regions	Cropping system	INM options
HP, J & K, Uttranchal, alluvial, sandy loam	Rice-mustard	Use of 40 kg N + 30 kg P_2O_5 + 40 kg K_2O + FYM/ GM @ 10 t/ha + *Azolla* @ 10 t/ha + 20 kg $ZnSO_4$ to rice crop and 20 kg N + 10 kg P_2O_5 + 25 kg K_2O to mustard crop (12).
-do-	Rice-potato	Applications of 40 kg N + 20 kg P_2O_5 + 15 kg K_2O + *Azolla*/GM @ 10 t/ha + 20 kg $ZnSO_4$ to rice and 50 kg N + 50 kg P_2O_5 + 30 kg K_2O + FYM @ 10 t/ha + seed treatment with *Azotobacter* and PSB to potato crop gave highest yield (24).
Western UP,alluvial with neutral alkaline reactions	Sugarcane-potato	• In autumn planted sugarcane, use of 100 kg N + 45 kg P_2O_5 + 30 kg sulphitation pressmud/ GM + incorporation of potato foliage (24). • In potato, application of 135 kg N + 20 kg P_2O_5 + 60 kg K_2O + FYM @ 10 t/ha + seed treatment with *Azotobacter* + PSB (24).
SK Nagar (Gujarat), sandy loam, arid	Pearl millet-wheat	50% NPK + 50% organic manure (FYM) to rainy season pearl millet and 100% fertilizer to winter wheat (24).

Contd.

Table 7.4 INM options for miscellaneous cropping systems *(Contd.)*

Regions/location soil type/ soil order ecosystem/agro-climatic regions	Cropping system	INM options
Hisar (Haryana), sandy loam, arid	Pearl millet-wheat	75% RDF through fertilizer + 25% N through FYM to kharif crop and 75% RDF through fertilizer to winter wheat (35).
Modipuram (UP), sandy clay, semi-arid	Sorghum-gram	Incorporation of FYM @ 6 t/ha + 50% RDF to sorghum as well gram proved effective (26).
Kanpur (UP), sandy loam, semi-arid	Groundnut-wheat	20:30:45 (N:P:K) + FYM @ 10 t/ha + gypsum @ 300 kg/ha to groundnut and 80:40 (N:P) to wheat crop (24).
Sangli (Maharashtra), medium black, semi-arid	Groundnut-wheat	Incorporation of 2.5 t FYM + 75% recommended fertilizers to kharif groundnut and 100% NPK to wheat crop (12).
-do-	Sugarcane-wheat	In sugarcane 135 kg N + 45 kg P_2O_5 + 30 kg K_2O + (FYM + sulphitation pressmud)/GM (*Sesbania*/ *sunhemp*) @ 10 t/ha and 80 kg N + 40 kg P_2O_5 + 40 kg K_2O to wheat sustained soil and crop productivity (24).
Rajasthan, MP, alluvial, red and black soils	Rice-gram	Application of 25 kg N + 15 kg K_2O + pulse crop residues incorporation + BGA @ 10 kg/ha/ *Azolla* @ 10 t/ha to rice and 10 kg N + 20 kg P_2O_5 + *Rhizobium* + 5 t FYM/ ha + 500 g PSB to gram improved yield and soil health (12).
Coastal plains and ghats, soils are lateritic, black and coastal alluvial	Pearl millet-wheat	Combine use of 25 kg N + 20 kg P_2O_5 + 10 kg K_2O + *Azotobacter*/ *zospirillum* to pearl millet and 45 kg N + 20 kg P_2O_5 + 15 kg K_2O + FYM @ 5 t/ha to wheat increased yield (24).

Contd.

Table 7.4 INM options for miscellaneous cropping systems *(Contd.)*

Regions/location soil type/ soil order ecosystem/agro-climatic regions	**Cropping system**	**INM options**
Udaipur (Rajasthan), sandy loam, semi-arid	Sorghum-chickpea	• Combined application of *Azospirillum* + *Azotobacter* and PSB biofertilizers with 5 t/ha organic manure at 20 and 50% reduced level of N and P supply from recommended (80 kg N and 40 kg P_2O_5/ha) can provide at least 25% higher grain yield of sorghum that achieved with 100% RDF (26). • In chickpea application of *Rhizobium* + PSB and 5 t organic manure/ha at 50% reduced level of N and P fertilizer application from the recom-mended (20 kg N and 40 kg P_2O_5/ha) can provide 60% higher grain yield of chickpea from achieved with 100% RDF (12).
Udaipur (Rajasthan), sandy loam, semi-arid	Soybean-winter maize	Combined application of *Rhizobium* + PSB + 5 t organic manure/ha @ 80% RD of N (20 kg/ha) and 50% of RD of P (40 kg/ha P_2O_5) in soybean and *Azospirillum* + *Azotobacter* and PSB with 5 t/ha organic manure at 80% of the recommended N dose (150 kg/ha) and 50% of the recommended P dose (160 kg P_2O_5/ha) was found most promising for reco-mmendation (12).
Pantnagar, medium alkaline soil (Uttranchal)	Sugarcane-sugarcane-wheat/Lentil/ Mentha	Application of combination of *Azotobacter diasotrophicus*, *Azospirillum*, compensate can yield due to 30–40% less application of fertilizer nitrogen (150 kg N/ha) + recommended nutrient of K_2O and P_2O_5 (24).

Contd.

Table 7.4 INM options for miscellaneous cropping systems *(Contd.)*

Regions/location soil type/ soil order ecosystem/agro-climatic regions	Cropping system	INM options
Saurashtra (Gujarat)	Groundnut- pearl millet	Use of PSM strain at 50% reduced level of P as DAP (50 kg P_2O_5/ha for ground nut and 40 kg P_2O_5/ha for pearl millet) shall have similar yield that obtained under 100% recommended dose of P fertilizer (12).
Vidarpha (MS), clay loam soil	Summer mungbean-winter wheat-pearl millet-chickpea	Organic manure @ 5 t/ha + 20 kg N + 40 kg P_2O_5/ha and brady-rhizobium + PSB appeared to the best package for mungbean and wheat. Wheat yield was satis-factory at par with that of 75% RDF (100 kg N + 60 kg P_2O_5/ha) best package (12).
Delhi, alluvial, medium alkaline	Rice-wheat-mungbean	Combined application of BGA and *Azolla* @ 60 kg/ha N supply from chemical fertilizer improve the nitrogen economy of kharif HYV rice in the specified soil locations provides yield at par with 90–120 kg N supply/ha as chemical fertilizer. Combined application of BGA + PSB reduced 30 kg P_2O_5/ha applied as single super phosphate in the recommended schedule of chemical fertilizers without any yield loss (12).

7.10 KEY REFERENCES AND RESOURCES FOR FURTHER READING

1. Abusaleha and Shanmugavelu, KG. Effect of organic vs inorganic sources of nitrogen on growth, yield and quality of okra. *Indian Journal of Horticulture* 1988; 29:312–318.
2. Acharya CL, Subbarao A, Biswas AK, *et al*. Methodologies and Package of Practices on Improved Fertilizer use Efficiency under Various Agroclimatic Regions for Different Crops/Cropping Systems and Soil Conditions. Indian Institute of Soil Science, Bhopal 2003; 38:p1–74.
3. Alam A (2003). Annual Report, University of Agricultural Sciences and Technology of Kashmir, Shalimar, Srinagar, 191 121, India.

4. Ariyarathe (2000). Integrated Plant Nutrition System (IPNS). Training Manual (Sri Lanka).
5. Balanced fertilizer use for crops on http:11www.indiaagronet.com/indiaagronet/ Technology – Upd/contents/integrated-nutrient. 10/30/2003.
6. Bhardwaj KKR. Role of Organic Waste in Soil Fertility Management. In: 50 years of Natural Resource Management Research (Singh G.B. and Sharma BR, Eds) NRM Division, ICAR, New Delhi 1998; p213–220.
7. Bhattachary P, and Mishra UC (1995). A book on biofertilizer for extension workers published by the National Biofertilizer Development Centre, Ghaziabad.
8. Doran JW, Jones AJ, Arshad MA. Determinants of Soil Quality and Health. In: Soil Quality and Health. In soil quality and soil erosion (Lal R Ed.), CRC Press, New York, USA 1999; p17–36.
9. Easwaran S, Kumar N, Marimuthu S. Studies on the Effect of Integrated Nutrient Management on Leaf Nutrient Status and Enzymes Activities in Tea (Camellia spp.). *Indian Journal of Horticulture* 2006; 63:224–226.
10. Gicana R Norlito (2001). Integrated Plant Nutrient Management in the Phillipines. In proceedings of Regional Workshop on IPNS development in rural poverty alleviation, Bangkok, Thailand.
11. Harmson K. Integrated Phosphorus Management. In integrated plant nutrition systems. FAO fertilizer and plant nutrition bulletin 1995; 12:p293–306.
12. Hegde DM, Dwivedi BS. Integrated Nutrient Supply and Management as a Strategy to Meet Nutrient Demand. *Fertilizer News* 1993; 38:49–59.
13. Integrated Nutrient Management: Role of Biofertilizer, Recommendations of DBT Network Programme. Department of Biotechnology, Ministry of Science and Technology, Government of India 1999–2003; p1–18.
14. Integrated plant nutrition system (IPNS) compendium. Published by Economic and social commission for Asia and the pacific, Fertilizer Advisory, Development and Information Network for Asia and the Pacific (FADINAP) 2002; p1–53.
15. Jose D, Shanmugavelu KG, Thamuraj S. Studies on the Efficacy of Organic vs Inorganic form of Nitrogen in Brinjal. *Indian Journal of Horticulture* 1998; 27:100–103.
16. Lampe Siegfried (1999). Principles of Integrated plant Nutrition Management System. FADINAP | FPA workshop on development of integrated plant nutrition system for the Philippines, Ternate, Carite, Phillippines.
17. Marwaha BC. Biofetilizers—A Supplementary Source of Plant Nutrient. Fertilizer News 1995; 40:39–50.
18. Marwaha BC. Scope and Significance of Fertigation in Indian Agriculture. *Fertilizer News* 2003; 48:97–101.
19. Motsara MR, Bisoyi RN (2001). 50 crop demonstrations on biofertilizers. National Biofertilizer Development Centre, Ghaziabad (UP) India.
20. Motsara MR, Bhattacharayya P, Srivastava Beena (1995). Biofertilizer Technology, Marketing and Usage, A source Book cum Glossary, FDCO Publication, New Delhi.
21. Nath B, Korla BN. Studies on Effect of Biofertilizers in Ginger. *Indian Journal of Horticulture* 2000; 57:168–171.

22. NBSSLUP (2004). Soil Resource Management Report. National Bureau of Soil Survey and Land Use Planning, Nagpur.
23. Parr JF, Papendick RI, Hornich SB, *et al.* Soil Quality: Attributes and Relationship to Alternative and Sustainable Agriculture. *American Journal Alternative Agriculture* 1992; 7:5–11.
24. Raghavaiah CV, Verma VS, More SD, *et al.* Oilseed Production in Salt Affected Soils. 2003; p40.
25. Sharma PD and Biswas PP. IPNS Packages for Dominant Cropping System in different Agro-climatic Regions of the Country. *Fertilizer News* 2004; 49:p43–47.
26. Sharma RP, Rana DS. Nutrient Management in Vegetable Crops for Sustainable Production. Fertilizer News 1993; 38:31–44.
27. Singh AK, Narayansamy D, Rattan RK, *et al.* Abstracts, National Seminar on Development in Soil Science, IISS, New Delhi 2001; p245.
28. Singh AK. Mantra for the New Millennium. *The Hindu Survey of Indian Agriculture* 2006; p32–40.
29. Singh K, Gill IS, Verma DP. Studies on Poultry Manure in Relation to Vegetable Production Cauliflower. *Indian Journal of Horticulture* 1970; 20:537–547.
30. Singh K, Minhas MS, Srivastava DP. Studies on Poultry Manure in Relation to Vegetable Production (potato). *Indian Journal of Agricultural Research* 1969; 26:69–73.
31. Singh SR, Sant P, Kumar J. Organic Farming Technologies for Sustainable Vegetable Production in Himachal Pradesh. *Himachal Journal of Agriculture Research* 2000; 26:69–73.
32. Somani LL, Totawat KL (2001). Integrated nutrient management for maintaining soil and crop productivity in Souvenir of the 66th annual convention of the Indian Society of Soil Science (Somani LL, Totawat KL eds) organized by Department of Agricultural Chemistry and Soil Science, Rajasthan College of Agriculture, Udaipur.
33. Srinivasarao Ch, Prasad JVNS, Vittal KPR, *et al.* Role of Optimum Plant Nutrition in Drought Management in Rainfed Agriculture, *Fertilizer News* 2003; p105–114.
34. Subbarao A, Chand Subhash, Srivastava S. Opportunities for Integrated Plant Nutrient Supply System for Crops/Cropping System in different Agro-eco-regions. *Fertilizer News* 2002; 47:75–90.
35. Chand Subhash (2001). Integrated nutrient management in mustard [*Braccica juncea* L. (Zern & Coss.)] Ph.D. thesis, MPUAT, Udaipur, Rajasthan.
36. Chand Subhash (Ed). Integrated Nutrient Management for Sustaining Crop Productivity and Soil Health, I. B. D. Co. Lucknow 2008; p112.
37. Chand Subhash, Mir AH. Microorganisms as a Component for Improving Phosphate use Efficiency by Crops. *Journal of Environment and Ecology* 2007; 23:p186–189.
38. Chand Subhash, Pabbi S. Organic Farming — a Rising Concept. In souvenir of Agriculture summit – 2005 organised by Ministry of Agriculture, Government of India and FICCI, Vigyan Bhavan, New Delhi 2005; p1–8.
39. Chand Subhash, Somani LL. Balanced Use of Fertilizer, Organics and Biofertilizer for Improving Yield of Mustard. *International Journal of Tropical Agriculture* 2003; 21:133–140.

40. Chand Subhash, Somani LL. Exploring Possibilities of Improving the Yield of Mustard [*Brassica juncea* L.) (Czern & Coss)] through Integrated Nutrient Management. *International Journal of Tropical Agriculture* 2005; 23:177–182.
41. Chand Subhash, Sahi NC, Ali T. Vermicomposting in Organic Farming. In Souvenir of Agriculture Summit-2006 organised by Ministry of Agriculture, Govt. of India and FICCI, Vigyan Bhavan, New Delhi 2006; p1–4.
42. Tewatia RK, Kawle SP, Chaudhary RS. *Indian J Fert* 2008; 3:1:111–118.
43. Tiwari KN, Sulewski Garin, Portch Sam. Challenges of Meeting Nutrient Needs in Organic Farming. *Indian Journal of Fertilizers* 2005; 4:41–48; 51–59.
44. Tondon HLS. In: Plant Nutrient Needs, Supply, Efficiency and Policy Issues (Kanwar JS, Katyal JC Eds) 1996; p15–28. National Academy of Agricultural Sciences, New Delhi, India.
45. Wani SA, Dar GH, Singh G, *et al.* (2003). Souvenir of the national seminar on organic products and their future prospectus. SKUAST-K, Shalimar Campus, Srinagar. 1–52.
46. Whitbread M, Antony Blair, J Graeme, *et al.* Managing Legume Leys, Residues and Fertilizers to Enhance the Sustainability of Wheat Cropping System in Australia. Soil physical fertility and carbon. *Soil and Tillage Research* 2000; 54:57–89.
47. Yadav DS, Alok Kumar. System based Integrated Nutrient Management for Sustainable Crop Production in Uttar Pradesh. *Fertilizer News* 1993; 38:45–51.
48. Yadav RL. Improving Nitrogen Efficiency. The Hindu Survey of Indian Agriculture 2006; p108.
49. Chand S, Singh L, and Singh P. Sustainable Agriculture, food security and climate change. Daya Publisher, N. Delhi, 2012: p218.
50. Chand S. Terminology of Soil Fertilitiy, Fertiliser and Organics. Astral International, 2014; p179.

INM is the best policy for higher yield and soil health. —Subhash Chand

CHAPTER 8 Opportunities for Precision Farming (PF) and Remote Sensing (RS) for Sustainable Soil and Crop Management

Precise application of organic inputs to farming leads eco-friendly ecosystem.
—Tej Pratap

The agriculture is changing day to day in modern scenario. Several new concepts, technologies and innovations have transformed the face of agriculture sciences. Resource conserving technologies (RCTs), precision farming (PF), and organic farming are examples of widely discussed and accepted technologies. Precision agriculture is an agricultural system that has the potential of dramatically changing agriculture in this 21st century. It is based on information technology, which enables the producer to collect information and data for better decision making. The major components of precision farming are: GIS, GPS/DGPS, remote sensing and farmer/variable rate applicator. Precision agriculture precisely assess the infield variability and manages the same for enhanced productivity and environmental safety through precision land leveling for managing landscape variability, site specific nutrient, site-specific planting, variable rate technology and other input application. The information for the variability may be obtained from soil tests, for nutrient availability, yield monitors for crop yield, soil samples for organic matter content, information in soil maps, or ground conductivity meters for soil moisture. The crop inputs are distributed on a spatially selective basis through grid sampling or management zone approach. In the present modern agriculture, precision agriculture have an important role in management of plant nutrient, water application, leveling of lands, control of diseases and pests. This chapter deals with all about meaning and principle, components, uses, variability in the fields, limitations and methodology of precision agriculture.

8.1 PREAMBLE

Agriculture is a very important sector for the sustained growth of the Indian economy. About 70 per cent of the rural households and 8 per cent of urban households are still principally dependent on agriculture for employment. Green revolution succeeded in India to increase the farmer's income, yield of major crops and made India self-reliant in food production, with the introduction of high yielding varieties and use of synthetic fertilizers and pesticides (Ghosh *et al.*, 1999).

In the post-green revolution period agricultural production has become stagnant and horizontal expansion of cultivable lands becomes limited due to burgeoning population and industralization. In 1952, India has 0.33 ha of available land per capita, which is now 0.15 ha (Singh *et al.*, 2000). As a result of increase in production through vertical dimension (yield per hectare basis) as horizontal becomes constrained, our agriculture becomes chemicalized through heavy use of synthetic fertilizers and pesticides which threatened the sustainability of our agriculture. In this situation, it is essential to develop ecofriendly technologies and concepts for maintaining crop productivity. One such technology and concept is precision farming.

Geographically, India is widely distributed into different agroclimatic zones, and the information needs for the farming system in these areas is entirely different. Though widely adopted in developed countries, the adoption of precision farming in India is yet to take a firm ground primarily due to its unique pattern of land holdings, poor infrastructure, lack of farmer's inclination to take risk, socioeconomic and demographic conditions (Mishra *et al.*, 2008). The PF in India could be unique in nature, different from that is practised in developed countries. It would be knowledge intensive with less involvement of sophisticated technologies. The work has already begun in this direction, with the initiation of PF research in many research institutes of international repute like ISRO, Ahmedabad, MS Swaminathan Foundation, Chennai, IARI, New Delhi, Project Directorate of Cropping Systems Research, Modipuram, etc. (Narayan, 2000 and Baburao, 2004).

8.2 PRECISION FARMING

"Art and science of utilizing advanced technologies for enhancing crop yield while minimizing potential environmental threat to the planet."

Precision farming is precisely assessing the infield variability and managing the same for enhanced productivity and environmental safety. This technology recognizes the inherent spatial variability that is associated with most fields under crop production. Once the infield variability is recognized, located, quantified and recorded, it can then be managed by applying farm inputs in specific amounts, at specific time and at specific locations. Khosla R (2001) Precision literally means "accuracy and it denotes degree of refinement in measurement. But in science it markedly differs from accuracy. Precision farming technology recognizes the inherent spatial variability that is associated with most fields under crop production (Thrikwala *et al.*, 1999).

Table 8.1 The difference between accuracy and precision

Accuracy	Precision
Signifies closeness to true value	Signifies closeness to the average value
Implies the lack of bias	Involves control of variability

8.2.1 Precision Farming —Why?

The green revolution has although increased productivity, but it has also resulted in several negative ecological impacts, such as:

- Depletion of lands
- Decline in soil fertility
- Soil salinization
- Soil erosion
- Deterioration of environment
- Health hazards
- Poor sustainability of agricultural lands
- Decrease in biodiversity, etc.

Besides, these issues, declining use efficiency of inputs and dwindling out-input ratio have rendered crop production less remunerative. Therefore, agricultural research seeks the generation of new technologies to reorient the current and future needs and constraints. The concept of precision farming may be appropriate to serve these problems though it looks unsuitable to Indian conditions, but is not impossible to adopt. The MS Swaminathan Research foundation, Chennai, India, has joined hands with Israel to initiate PF on an experimental basis including conducting training programmes (Narayan, 2000).

The environmental risks provide a basis for examining agriculture's impact on environmental quality and the magnitude of that impact (Table 8.2). Schepers (2000) suggested that hypoxia a condition where water is depleted of its oxygen content, which results in a serious reduction of biological activity, e.g. Gulf of Mexico where 7,000 miles sq area has developed into a hypoxic zone and is believed to be caused by an abundance of nitrogen, phosphorous, and silicate in the water that could result from runoff, erosion, subsurface drainage, and base flow from field into adjacent streams. These transport processes, when coupled with nutrients, heavy metals, pesticides, antibiotics, pathogens and sediments, give rise to different movement of these containments across a watershed.

Table 8.2 Environmental risks from nutrients and soil organic matter

Process	N	P	K	S	OM
Leaching	+	0	–	–	–
Denitrification	+	–	–	–	–
Eutrophication	+	+	–	–	–
Precipitation	+	+	+	–	–
Run off	+	+	0	0	+
Volatilization	+	–	–	0	–
Saltation	–	–	+	–	–

Source: Hatfield, 2000

The process outlined in the Table 8.2. cannot be changed, but it is possible to modify the loading of nutrients and pesticides in a field. Nutrients can be applied as inorganic or organic sources to a field, and the timing of application may be an effective method of avoiding excessive nutrients or pesticides when transport mechanisms are favorable for off-site movement.

8.2.2 Variabilities in Field

- Land is heterogeneous
- Soil fertility status, i.e. nutrient content varies from one place to another even in the same field
- Population of weeds is never uniform throughout the crop field
- Population dynamics of insects-pest are variable in nature
- Disease intensity may not be same throughout the field
- Water requirement also may vary within the field

Precision agriculture emphasises on this aspect and deals with judicious crop management practices at micro-level wherein only required amount inputs are applied. It is a form of farming where site specific management practices are adopted giving due consideration to the spatial variability of land in order to maximize crop production and minimize environment damage. In conventional agriculture without considering these variable fertilizers, herbicides, insecticides, fungicides, etc. are applied at a uniform rate throughout the crop field.

8.2.3 Components of Precision Farming

The major components of precision farming are:

i. GIS
ii. GPS
iii. Remote sensing
iv. Farmer

GIS (geographic information system)

It is a system of capturing, storing, checking, integrating, manipulating, analysing and displaying data which are spatially referenced to earth.

Components of GIS

- Computer system and software,
- Use spatially referenced or geographical data, and
- Management and data analysis.

i. **Computer system and software:** GIS runs on the whole spectrum of computers systems ranging from portable personal computers (PCS) to multi-user super computers, and are programmed in a wide variety of software languages. There are number of elements that are essential for effective operation of GIS which are as follows:

a. Process or with adequate power to run the software
b. A sufficient memory for data storage
c. A good quality with high resolution color graphics screen
d. Input and output devices for data, e.g. scanners, keyboard, printer, etc.

ii. **Spatial Data:** Spatial data are characterised by information about position, connections with other features and details of non-spatial characteristics. The spatial referencing of spatial data is important and should be considered at the out set of any GIS project. Spatial data representing layers/objects must be simplified before they can be stored in the computer.

iii. **Management and analysis of data:** A GIS should be able to perform data input, storage, management, transformation, analysis and output. Data input is the process of converting data from its existing form to one that can be used by the GIS. It needs to handle two types of data — graphical data and non-spatial attribute data. The graphical data describes the spatial characteristics of the real world. Non-spatial attribute data describes with the features represent. Transformation and the procedure for analysis can also be classified based on the amount of data (Das, 2007).

8.2.4 Strategies for Implementation

The method of implementation is one of the important factors that can affect the success or failure of GIS programme and this falls into four main categories. These are:

- Direct conversion, from the old system to new
- Parallel conversion, where both old and new system run along side one another for a short time period
- Phased conversion, where some of the functions of the old system are implemented first, then others follow, and
- Trial and dissemination, i.e. conversion to a new system

(Das, 2007)

8.2.5 Techniques for Identification of GIS Problem

There are two techniques that are used for the identification of the GIS problem.

i. Creating a rich picture, and
ii. Developing a root definition.

Creating a rich picture: It is a schematic view of the problem a project will address. It represents main components of the problem, as well as any interactions that exist. The development of a rich picture should not be rushed, particularly if it is trying to reflect an unsaturated problem. A poorly defined rich picture may translate into a poor GIS application.

Root definition: It is the view of a problem from a specific perspective. Establishing a common root definition for a problem with the help of others to evaluate and understand why a GIS has been constructed in a particular way.

Once rich picture and root definition exist main aims and objectives for a project can be identified and a GIS data model can be created (Das, 2007).

8.2.6 Problems for Implementation

- Data in the wrong format for the GIS software,
- Lack of GIS knowledge imposing technical and conceptual constraints on a project, and
- Users of the GIS frequently changing their mind about what they want the GIS to do

(Das, 2007)

8.2.7 Success of GIS

- Better tools for the handling and exchange of information,
- A better served GIS community (with conference, text books, magazines, etc.),
- New players in the GIS industry,
- The development of standards for data issues, and
- The development of a core set of ideas to inform teaching and training in GIS

8.2.8 Limitations of GIS

- Two dimensional with limited abilities to handle the third dimension,
- Static, with limited abilities to cope with temporal data,
- Good at capturing the physical position of objects, their attributes and their spatial relationships, but with very limited capabilities for representing other forms of interaction between objects,
- Offerings a diverse and confusing set of data models, and
- Still dominated by the idea of map, or the view of spatial database as a collection of digital maps

(Goodchild, 1995)

8.2.9 Miscellaneous Uses

- GIS is the principle technology used to integrate spatial data coming from various sources in a computer (Heimlich, 1998).
- GIS technique deals with the management of spatial information of soil properties, cropping system, pest infestations and weather conditions.
- This is primarily an intermediate step because it combines the data collected at different times based on sampling regimes, to develop the subsequent decision technologies, such as process model, expert system, etc.

8.2.10 Applications of GIS

- GIS techniques make weed control, pest control and fertilizer application site-specific, precise and effective, it would also be very useful for drought monitor-

ing, yield estimation, pest infestation monitoring and forecasting (Singh, 2000; Reddy and Anand, 2000 and Biswas and Subba Rao, 2000).
- GIS coupled with GPS, microcomputers, RS and sensors is used for soil mapping, crop stress, yield mapping, estimation of soil organic and available nutrients (Biswas and Subba Rao, 2000).
- In combination these technologies have brought out rapid changes in data collection, storing, processing, analysis and developing models for input parameters (Reddy and Anand, 2000).

8.3 GLOBAL POSITIONING SYSTEM (GPS)

The GPS was developed by the US military and later permitted for restricted civilian use. GPS receivers provide a method for determining location anywhere on the earth. Accurate, automated position tracking, with GPS receivers allow farmers and agricultural service provides to automatically record data and apply variable rates of inputs to smaller areas within larger fields (Donal and William, 1998). The inherent accuracy of GPS is about 5 m, which is based on 95 per cent probability that the position given will be within 5 m of the true value position.

The GPS can be used in two modes; single receiver mode and differential mode (DGPS) using two receivers. Single receiver collects the timing information and processes it into position. This system is the cheapest and easiest, but its accuracy suffers due to the introduced positional error. In the differential mode (DGPS), one receiver is mounted in a stationary position, usually at farm office while the other is on the machine/implement (Yadav and Subba Rao, 2000).

8.3.1 Systems of GPS (global navigation satellite system)

- The NAVSTAR global positioning system (GPS) is owned by the Government of the USA, and
- The global navigation satellite system (GLONASS) is controlled by a consortium headed by the Russian Government.

Both systems are built using a space segment comprising a constellation of dedicated satellites, a control segment that monitor, maneuvers and updates information to the satellites, and a user segment trying to determine accurate ground position (ACPA, 1999).

8.3.2 GPS Operations

- In basic terms a GPS users position is determined by resection, using the distances measured to the satellites. There are three techniques for calculating these distances based on information provided on two transmission frequencies from the satellites. These techniques are known as:
- C/A code (coarse/acquisition),
- P code (precision), and
- Codeless

The C/A code technique is also known as the standard positioning service (SPS) and is available for all civilian use and is most commonly used in PA.

The P-code technique is reserved for military use and is also known as precise positioning service (PPS).

The codeless technique requires more sophisticated and expensive receivers.

With C/A code techniques distances are estimated by measuring the travel time of a codal signal from each satellite and multiplying it by the transmission velocity (the speed of light). The information coded into the signal includes satellite orbit, current position and time information. The time information is provided by four extremely accurate atomic clocks onboard the satellite.

The distance of 4 satellites must be instantaneously determined by users receiver (remote receiver) in order to obtain a point position in latitudes, longitudes and elevation. Each satellite is required for resolving latitude, longitude and elevation and 4th is required to determine errors between the satellite receiver time pieces (0.07 sec) (ACPA, 1999).

8.3.3 How GPS Works?

Figure 8.1 shows the important considerations in GPS technology. It uses a space-age update to the same principle of triangulation that we learned in high school geometry. One of the satellites sends a signal towards each, stating the exact time. But when a GPS receiver on the ground checks the time, it is little off. The time lag times the speed of light that the radio waves travel tells us how far away the satellite is. Knowing the position and distance to a set of satellites allows calculation of the position of GPS receiver.

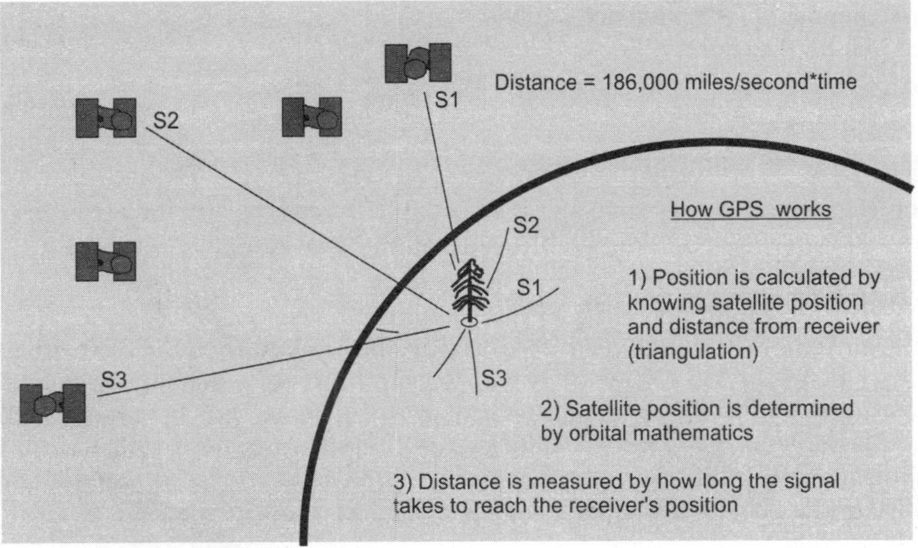

Fig. 8.1 Conceptual framework for GPS

Although the process involves complicated electronics, it uses the same calculation we use in geometry class involving that device with a pencil on one arm and a life-threatening sharp point on the other. Recall that we would stock the piece of paper then extend the arms for a given distance (satellite to GPS receiver time lags × speed) and make a small arc. Repeat for a second point/distance and where the arcs cross determines where we are in two dimensional spaces.

In 3-dimensional space, circles of calculated distance are mathematically "drawn" about a set of satellites whose precise position is known at any instant in time through orbit mathematics. The intersection of the circle determines the position of the GPS receiver on the earth (Racher, 1999).

8.3.4 Cost Estimation of DGPS Network Infrastructure Development in India

As a first step towards operationalization of PF, it is necessary to establish a DGPS infrastructure for the country. A DGPS network would cater to the needs of multitude of applications (meterology, transportation, geodetic survey, crystal deformation studies, disaster management and mitigation, etc.) of which PF is one? Each DGPS master reference is capable of providing the services within a radius of 100 kilometre radius. Hence, the number of master stations required to establish the DGPS infrastructure, which would cover the entire country, was calculated by Mishra *et al.* (2008) and is given in Table 8.3.

Table 8.3 Cost estimation of DGPs network in India

Area of India	= 329 million hectares
Area of GPS (circular area, PI = 3.14)	= PI* 1002 sq km
	= 31400 sq km
Total number of GPS reference stations required for the country	= 329/3.14
Total GPS reference stations required	= 105
Cost of a single DGPS set	= ₹ 3 lakhs
Total cost of the entire infrastructure	= ₹ 3.15 crores

The calculation shows that this is affordable by any means for a country as a whole considering the innumerable application it can cater.

8.4 REMOTE SENSING (RS)

The term remote sensing was introduced in the USA in the late 1950s to attract funding from the US Office of Naval Research. Remote sensing is defined as "covering the collection of data about objects which are not in contact with the collecting device". Remote sensing is a multidisciplinary activity which deals with the inventory, monitoring and assessment of natural resources through the analysis of data obtained by observations from a remote platform.

Remote sensing devices are generally are of two types:

Active sensing: When the sensing device direct EMR at an object and detects the amount of that energy which is reflected back, e.g. radar, radiowave emissions (wavelength >10^3 μm) and laser-imaging radar (UV, visible and near infrared, 10^{-3}–10 μm).

Passive sensing: When the sensing device detects EMR originating from another source, primarily from sun.

8.4.1 Types of Remote Sensing

Based on distinction regarding the collection of data:

i. Air borne remote sensing: Aerial photography is very useful to interpret and map soils, but the interpretation must be done very carefully because of risks or error. The value of black and white infrared aerial photographs was well recognized for separating stands of conifers (gymnosperms) and broad leaved species (hardwoods/angiosperms) in the cool temperature natural forests. However, this technique was not suitable for separating native conifers from eucalyptus. Black and white infrared aerial photography was widely accepted as the most suitable film type for mapping the boundaries of H_2O surfaces and for separating wet and dry lands.

Natural color photography is widely recognized for its role in disease detection and identification of some tree species growing in temperate zones. Color infrared photography at large, medium and small is now increasingly used in periodic national inventories and combined with satellite derived data.

Aerial photography at very large scale (e.g. 1:1200) has been demonstrated to be suited to the reliable identification of trees of individual species in cool temperate forests and is used operationally in forest inventories.

Aerial photography is used for soil survey and the scale of aerial photography depends upon the flying height of the air craft above the ground and the focal length of the camera.

$$\text{Photo scale} = \frac{\text{Flying height}}{\text{Focal length}}$$

Aerial photography used for the identification of changes in the land surface patterns that may be relatable to differing soil properties. In addition, it is used for interpreting soil boundaries and planning field work, for field navigation and plotting.

ii. Satellite remote sensing: With the development of satellite technology, recently satellite data is being applied more and more in forestry, particularly in multitemporal and multistage surveys of forests resources. Multitemporal refers to the time-wise collection of data about a site on more than one occasion. Multistage infers that the data is collected simultaneously for the same time for more than one altitude. (UNCOPUOS, 2007)

8.4.2 Stages of Remote Sensing/Principles

- Origin of electromagnetic energy (e.g. sun, transmitter carried by the censor)
- Transmission of energy from the source to the surface of the earth and its subsequent interaction with intervening atmosphere
- Interaction between energy and earth surface
- Transmission of the emitted or reflected energy to the remote sensor
- Detection of the energy by the sensor converting into photographic image or electrical output
- Recording of the sensor output
- Data processing for the generation of database
- Collection of ground truth and other collateral information
- Data processing and interpretation

Table 8.4 Relevant remote sensing techniques and the attributes estimated

Observation technique	Platform	Attribute estimated	
		Soil	Crop
Visible/NIR reflectance	Aircraft/satellite	Moisture	Leaf area index
		Organic matter	Biomass
		Texture	N status
		Salinity	Photosynthetic activity
			Species identification
			Physical damage
Thermal infrared	Aircraft/satellite	Moisture	Canopy temperature
			Moisture stress
			Vigour
Radar	Aircraft/satellite	Moisture	Leaf area index
		Surface roughness	Biomass
			Surface roughness
Gamma emission	Aircraft	Mineralogy Clay content	

Above table summarises the remote sensing techniques and the relevant attributes that can be estimated. This form of data appears suitable to quantify more core scale variation, but as the resolution of the technology increases, and ground truthing is improved; this may become a more useful tool for assessing small scale variation.

Table 8.5 Options for continuous sensing of soil attribute variation

Soil attributes	Measurement technique
Texture	Visible and NIR reflectance Electromagnetic induction (EMI) Ground penetrating radar (GPR) Acoustic sensors Tillage draft
Moisture	Electromagnetic induction (EMI) Ground penetrating radar (GRP) Electrical resistance Electrical capacitance Time-domian reflectivity (TDR) NIR reflectance Nuclear magnetic resonance (NMR)
Organic matter	Visible and NIR reflectance
Nitrogen	Ion selective electrode Ion selective field effect transistor (ISFET) Electrical conductivity
Ph	Ion selective electrode Ion selective field effect transistor (ISFET)
Salinity	Electromagnetic induction (EMI)
Compaction	Penetrometer
Top soil depth	Electromagnetic induction (EMI) Ground penetrating radar (GPR)
Horizon boundaries/clay pans	Electromagnetic induction (EMI) Ground penetrating radar (GPR)

Figure 8.2 reflects the application of remote sensing. The physiology of a leaf determines the relative absorption and reflection of light. The equal portions of blue, green and red light from the sun are basically unaffected by the surface of the leaf, but when it encounters the chloroplasts containing chlorophyll a and b it is radially altered. These pigments absorb most of the blue and red light for the energy needed on photosynthesis used in plant growth and maintenance. Other pigments in the leaf (i.e. carotenes) absorb lesser amounts of the other wavelengths of light.

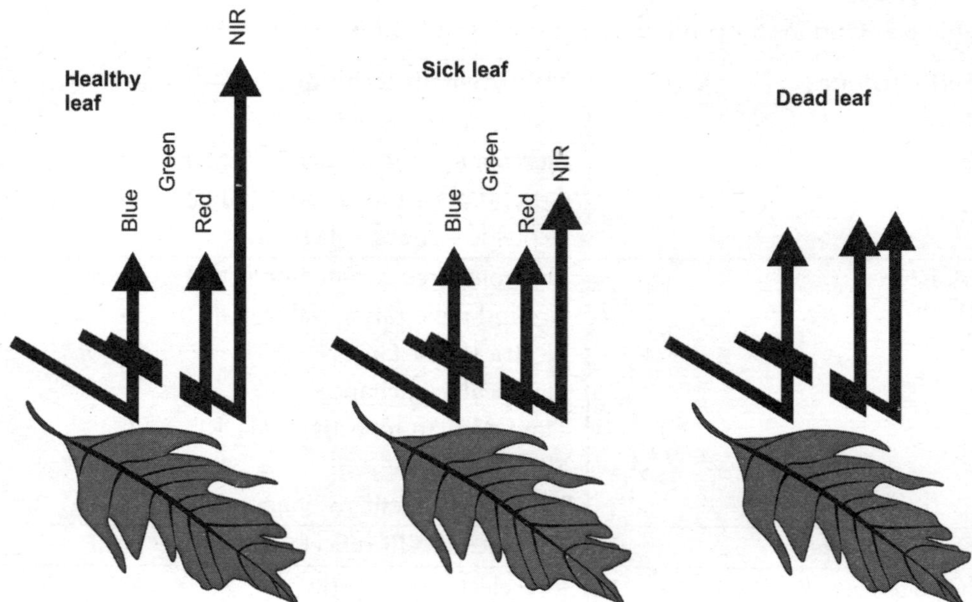

Fig. 8.2 a Plant physiology determines the quality of reflected light from a leaf

As the pigments altered light continues deeper into the leaf, it interacts with the spongy mesophyll. This bubble like structure like a mirror and reflects the light back towards the sky. Since the blue and red wavelength has been diminished, a predominance of green in the reflected light is seen — a healthy "green leaf".

An unhealthy leaf looks a lot different, particularly in remote sensing imagery. When water pressure changes, the spongy mesophyll in the leaves collapse with in hours and this area's efficiency of reflecting light is greatly reduced. The chloroplasts, on the other hand keep on working away at photosynthesis for several days. The result is that we "see" a slight change in reflectance (predominantly green) at first, then a slow progression to brown as the chloroplasts eventually quit preferentially absorbing blue and red light.

Remote sensing view differently reflected light beyond visible blue, green and red light. These wavelengths invisible near infrared light (INR) are unaffected by the plant's pigments and are highly reflected by the spongy mosophyll. When "bubbles" in this portion of a leaf collapse, there is an immediate and dramatic change in the reflectance of near infrared light (Racher, 1999).

It (Fig. 8.2b) extends the discussion of the basic concepts of plant physiology and its interaction with light from a plant to a whole field. From a simplified view, as more biomass is added the reflectance curve for bare soil (similar to a dead leaf) is transformed into a spectral signature that typifies one big green leaf. As the crop matures, the reflectance pattern changes again (Racher, 1999).

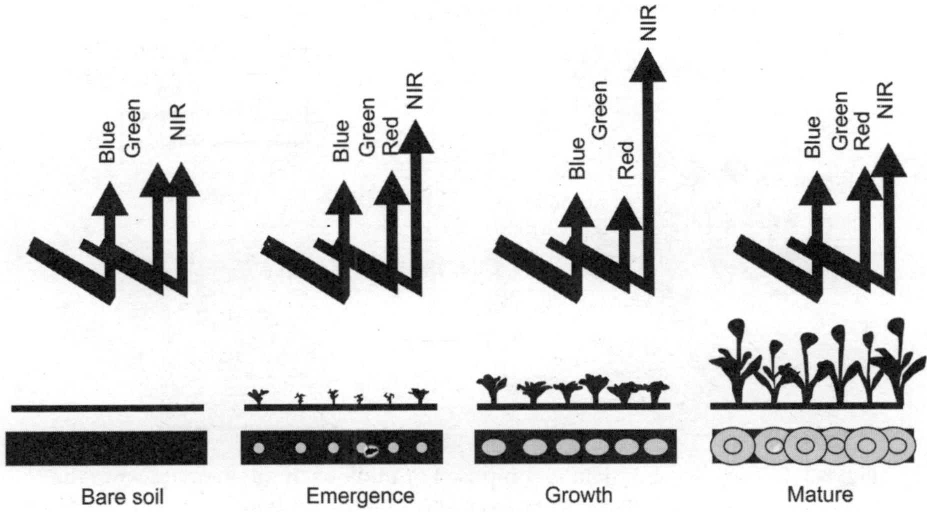

Fig. 8.2 b Reflectance from various field conditions

The scanner (Fig. 8.3) in a satellite operates a bit differently more like a laser printer that "sees" the world through thousands of dots. Its sensors focus for an instant at a spot on the ground (a few meters in diameter). Like our eyes, it records the relative amount of different types of light it "sees" — a lot of green for a dense healthy crop, much less green and more blue and red for bare ground. In addition to normal light (visible spectrum), it can record other types that we cannot see, such as near infrared, thermal and radar energy. The sensor sweeps from side to side and the satellite moves forward, recording the relative amounts of light reflected from millions of spots on the ground. When these spots (termed pixels for "picture elements") are displayed on a computer, they form an image similar to an aerial photography. In fact, a photograph can be "scanned" to generate a digital image like pulling the satellite out of the sky and passing it over the photo instead of the actual terrain. The important point is that behind any digital image there are millions of numbers recording the various types of light reflected from each instantaneous spot.

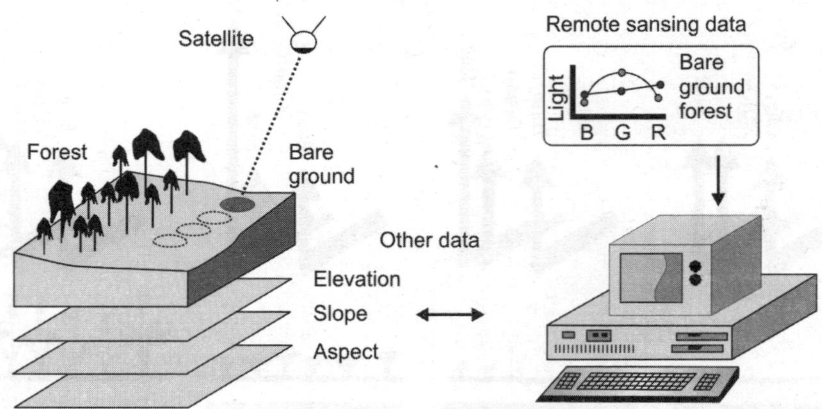

Fig. 8.3 Remote sensing data is composed of millions of numbers tracking the reflected energy from spots on the ground

Steps in precision farming:

- Assessing variability
- Managing variability

Assessing variability:

- Very crucial and first step of precision agriculture
- Spatial variability of all the determinants of crop yield should be well recognized, adequately quantified and properly located.
- Construction of condition maps on the basis of the variability is a critical component of precision farming.
- Condition maps can be generated through:
  i. Surveys,
  ii. Point sampling and interpretation,
  iii. Remote sensing (high resolution), and
  iv. Modelling

Managing variability:

- Precision land levelling
- Variable rate technology
- Site specific planting
- Site specific nutrient management
- Site specific weed management
- Precision pests and disease management
- Precision water management

8.5 PRECISION LAND LEVELING

- Levelness of land greatly influences all the farming operations
- Precision land leveling has been rated second only to development of high yielding varieties

Advantages of PLL

- It ensures higher resource use efficiency
- Soil moisture status of a field depends on levelness and influences yield and input use efficiency
- Significant improvement in crop productivity (5%–5%), water productivity (20%–30%), and nutrient use efficiency (15%–20%) under rice wheat cropping system in Indo-Gangetic plains has been reported due to precision land leveling (Jat *et al.*, 2004).
- Zero-till seedling was noted to perform better on well-leveled field as compared to unleveled or fairly leveled due to better seed placement, seed germination and uniform distribution of plant nutrients as well as irrigation water (Jat *et al.*, 2005).

Table 8.6 Agronomic efficiency (kg ha^{-1}) of N (AE-N), P (AE-P) and K (AE-K) under different land leveling systems in rice

Treatment	AE-N		AE-P		AE-K	
	2003	**2004**	**2003**	**2004**	**2003**	**2004**
LL + NPK*	18.75	20.00	86.54	92.31	56.25	60.00
TL + NPK*	7.65	9.17	35.38	42.31	23.00	27.50
TL + NPK #	–	–	–	–	–	–

LL– Laser leveling, TL–Traditional leveling, *N @ 120 kg, P@ 26 kg and K @ 40 kg ha^{-1} #– NPK (control)

It (Table 8.6) depicts the significant increase in the agronomic efficiency of the applied N, P and K in a typic ustocrepts in rice due to precision land leveling compared to traditional leveling. Improved use efficiency of the applied nutrients under laser land leveling is an obvious consequence (Jat *et al.*, 2004).

8.6 VARIABLE RATE TECHNOLOGY

- It is the technique of applying the farm inputs, such as seeds, fertilizers, pesticides, etc. in varying rates at the places and amounts they are required to produce uniform yields throughout the entire field.
- Variable rate seed drills and planters, fertilizer spreaders and sprayed are commercially available.
- Farm machineries equipped with VRT generally possess a DGPS receiver to locate the spatial variability in the field and automatically regulate the rate of application.

Disadvantages of VRT

- High cost of integrated control systems
- Operation needs technical know-how

8.7 SITE-SPECIFIC PLANTING

- Involves proper placement of seeds and maintaining optimum population to achieve maximum yield and quality.
- Known correlations exist among spatial variabilities in seed spacing, plant populations, yield and crop quality.
- Improper seed distribution and skips cause varied plant growth and increased competition between individual plants.
- Making spacing with 75% or more accuracy, the crop yield can be increased by 5%–10% and quality could be improved by 20%.

(Hess *et al.*, 1999)

Table 8.7 Comparative performance of precision and traditional seedling machines

Seedling methods	Grain yield (t ha^{-1})		Water productivity (kg grain m^3 water)	NUE (kg grain) kg^{-1} (N)
	On-station	On-farm		
Broadcasting (FP)	3.86	3.91	0.90	32.17
Traditional drill	4.10	4.37	0.95	34.17
Precision drill	4.84	4.76	1.13	40.33

The significant improvement in wheat yield, water productivity, and nitrogen use efficiency has been recorded with precision drill both under on farm station as well as on farm situations compared to farmers practice (broadcasting) and traditional drills (Pal *et al.*, 2004).

The investigation on assessing soil and yield variability in rice wheat system clearly demonstrated that yield variations in both the crops were mainly due to spatial variation in soil properties. Site-specific application has resulted in improvement in productivity of crops compared to blanket application across the entire field and also reduced the losses of nutrient (pal *et al.*, 2002).

8.8 SITE-SPECIFIC NUTRIENT MANAGEMENT (SSNM)

- Most important and widely used form of VRT is variable rate fertilizer application.
- It is well known that spatial variability in soil properties exists across the landscape.
- Uniform application of fertilizers results in under fertilization of certain parts of a field and over fertilization in other areas.
- Under fertilization results in yield loss and over-fertilization can be harmful to the crops and environment as well.

- There are significant correlations between yield and soil properties viz. slope, CEC, water holding capacity, etc.
- Spatial variability of P & K contents in the soil affects yield and tuber quality in potato.
- In a study in Australia, variable rate gypsum application on high sodic soils showed a benefit of \$563 ha^{-1} over five years as compared to uniform application.

(Cambouris *et al.*, 1999)

Table 8.8 Rice (PB-1) grain yield and N-use efficiency under different N management practices (average of 20 farmers)

Nitrogen management practices	Total N applied (kg ha^{-1})	Grain yield (ha^{-1})	Agronomic efficiency (kg ha^{-1} N)
No-N	0	2.75	–
Recommended nitrogen management	80	3.86	13.9
LCC <3 (No basal N)	80	4.18	17.9
80% basal + LCC <3	104	3.62	8.4
Farmers practice (No_3 splits)	100	3.74	9.9

Table 8.8 depicts the results demonstrations that were carried out on LCC based real N application in Basmati rice in Western UP and found significant improvement in yield and agronomic efficiency (AE_N) with LCC based N application than the fixed time N application (Singh *et al.*, 2005).

Table 8.9 shows the SSNM investigation carried out on hybrid rice-wheat cropping system at Modipurum. The SSNM package (N_{170}, P_{30}, K_{120}, S_{20}, Zn_7, Mn_{17}, $B_{0.6}$ in rice and N_{150}, P_{30}, K_{120} in wheat) resulted in significant yield advantages both in rice (9.95 t ha^{-1}) and wheat (5.94 t ha^{-1}) compared to soil test crop response (STCR), local ad hoc recommendation (LAR) and farmers practice (FP) (Shukla *et al.*, 2004). They also reported that the increase in net income (US \$ ha^{-1}) in SSNM was 471, 384 and 377, respectively over FP, STLR and LAR.

Table 8.9 Productivity and profitability of hybrid rice-wheat cropping system under site-specific nutrient management practices

Treatment	Grain yield (t ha^{-1})		Economics of the system (US\$ ha^{-1})			
	Rice	Wheat	Total net returns	Increase over FP	Increase over STLR	Increase over LAR
SSNM	9.95	5.94	1135	471	384	377
LAR	8.03	4.86	759	94	7	0
STCR	7.94	4.62	752	87	0	–7
FP	7.29	4.53	654	0	–87	–94

8.9 PRECISION WATER MANAGEMENT (PWM)

- Water is a critical and most limiting input for crop production.
- Crop yield generally shows linear correlation with amount of water transpired.
- When water is limiting factor, the crop yield is much lower than potential yield, whereas, excess watery may induce stresses in aeration and nutrient availability.
- Three approaches viz. (i) variable rate irrigation, (ii) soil — landscape water management, and (iii) drainage.

i. Variable Rate Irrigation

- Well-managed site-specific irrigation system dependent on physical circumstances and socio-economic conditions.
- Ensures optimum spatial and temporal distribution of water so as to promote the crop growth and yield.

ii. Soil-Landscape Water Management

- Variable rate irrigation alone cannot ensure precision watering due to differences in water availability which are governed by variation in landscape, different soil types and presence of soil degradation processes, such as erosion, compaction, salinisation, etc.
- It has been observed that the north facing slopes had 20 per cent more available water than south facing slopes.
- Crop yields are often higher in lower slopes due to greater availability of nutrients and soil moisture.
- Eroded soils as well as compacted soils are reported to have lower infiltration rates.
- Redistribution of water within a landscape due to either runoff or subsurface horizontal flow is also responsible for causing spatial variability in water availability.
- Laser land leveling, soil amelioration, etc. may be employed for landscape management.
- However, the agronomic inputs including irrigation water have to be applied in accordance with spatial variability in water availability within a field.

iii. Drainage

Practice of removing excess water from the field by surface or subsurface channels. Despite VR-irrigation, water may be found excess in some parts of the field due to soil-landscape variability.
(Jat *et al.*, 2005)

8.10 SITE-SPECIFIC WEED MANAGEMENT (SSWM)

- Degree of spatial variability exists in weed distribution.
- Over application of herbicides causes environmental contamination, increased cost and injury to the crops.
- Under application causes poor weed control with substantial yield losses.

- Spatial variability in weed density and species across a crop field is due to differential seed dispersal mechanisms, variation in physico-chemical properties and previous management practice. There are three approaches:

i. **Collection of data on weed population density and species and then preparation of digital application maps based on the population pattern.**

- VR-herbicide spraying is made corresponding to the application maps.
- 7%–69% reduction in herbicide requirement has been reported in maps, based on variable herbicide application.
- Sensor based spot spraying (it automatically detects the presence of weeds and applies herbicide accordingly).
- Consists of weed detection sensors mounted on spraying equipment.
- Sensors can distinguish targeted weeds from the crop and the information thus generated is passed to a central system that turns the sprayer on or off accordingly.
- Reduction of 47 per cent in herbicide cost while evaluation variable rate herbicide application in cereal crops has been noticed.
- Involves variable rate herbicide application according to soil physical and chemical properties.
- Herbicide rates are varied according to soil organic matter, soil structure, soil pH, etc.
- Variable application maps are developed on the basis of soil property maps.

(Sharma and Biswas, 2005)

8.11 PRECISION INSECT, PEST AND DISEASE MANAGEMENT

- Over application of pesticides leads to the problem of chemical residues in soil as well as in the produce.
- Application at sub-lethal doses may lead to development of resistance and resurgence in pets.
- In precision agriculture, doses of insecticides are to be decided on the basis of population dynamics of pests in management zones and disease management strategy is based on the disease incidence and disease severity across the landscape.
- Spatial variability of disease occurrence could well be taken into account while managing the diseases.
- Monitoring of insect population density is very difficult due to their highly mobile nature and lack of commerically available sampling scheme.
- Repeated sampling (insect population density) and increased costs.
- In precision pest management, a map of insect density will be more useful than a mean estimate of the same :

Two approaches:

i. Grid sampling
ii. Management zones

i. Grid Sampling

- Interpolated values are classified using GIS techniques into limited number of management zones.
- Boundaries of management zones are then visualised using mapping software, and management recommendations are developed for each zone.

ii. Management Zones

- More economically feasible approach.
- Divides fields into different regions referred to as production level management zones has been established recently to the US.
- Production level management zones (PLMZ) are defined as homogenous sub-regions of field having similar limiting factors.
- Delineation of management zones uses three GIS data layers viz. bare soil imagery, topography and farmers experience.
- A field can be divided into three different zones: high, medium and low based on the productivity of the area and the crop inputs can be applied accordingly.

(Sharma and Biswas, 2005)

8.12 PRACTICAL PROBLEMS IN INDIAN AGRICULTURE

Precision farming has been mostly confined to developed countries. Reasons of limitations of its implementation in developing countries like India are:

- Small land holdings
- Heterogenity of cropping system and market inperfections
- Lack of technical expertise knowledge and technology (India spends only 0.3% of its agriculture gross domestic product in research and development)
- High cost

(Mishra *et al.*, 2008)

8.13 METHODOLOGY TO BE ADOPTED

- Creation of multidisciplinary teams involving agricultural scientists in various fields, engineers, manufactures and economists to study the overall scope of precision agriculture.
- Formation of farmer's cooperative since many of the precision agriculture tools are costly (GIS, GPS, RS, etc.)
- Government legislation restraining farmers using indiscriminate farm inputs and thereby causing ecological/environmental imbalance would induce the farmer to go for alternative approach.
- Pilot study should be conducted on farmer's field to show the results of precision agriculture implementation.
- Creating awareness amongst farmers about consequences of applying imbalanced doses of farm input like irrigation, fertilizers, insecticides and pesticides.

8.14 ECONOMIC FEASIBILITY OF PRECISION FARMING IN INDIA ON AGRICULTURAL CONDITION

Unlike some new technologies, there is no clear answer as to whether or not PA is economically beneficial in the Indian agriculture condition.

- On one hand there are depletions of ecological foundations of the agro-systems, as reflected in teams of increased land degradation, depletion of water resources and rising trends of floods, drought and crop pests and diseases. On the other hand, there is imperative socio-economic need to have enhanced productivity per units of land, water and time.
- At present, 3 ha of rainfed areas produce cereals grain equivalent to that produced in 1 ha of irrigated. Out of 142 million ha, net sown areas, 92 million ha. are under rainfed agriculture in the country.
- From equity point of view, even the record agriculture production of more than 200 mt is unable to address food security issue. A close to 60 mt food grain in the storehouses of Food Corporation of India (FCI) is beyond the affordability and across to the poor and marginalized in many pockets of the country.
- Globally, there are challenges arising from the globalization especially the impact of WTO regime on small and marginalized farmers.
- Some other unforeseen challenges could be anticipated global warming scenario and its possible impact on diverse agro-ecosystems in terms of alterations in traditional crops.

(Baburao, 2004)

8.15 INDIANS' INITIATION AND CONCLUSIONS

The study on PF has been initiated in many research institutions, these are:

- ISRO (Ahmedabad) has started experiment in the Central Potato Research Farm at Jalandhar, Punjab to study the role of remote sensing, GIS and GPS in mapping the variability.
- MS Swaminathan Foundation, Chennai, in collaboration with NABARD has adopted a village in Dingugal district of Tamil Nadu for variable rate input application.
- IARI, New Delhi has drawn up plans to do precision agriculture experiments in the Institute's Farm.
- Project Directorate for Cropping System Research (Modipuram), and Meerut (UP) has initiated a project on precision agriculture in collaboration with Central Institute of Agricultural Engineering (CIAE), Bhopal.
- ISRO has also initiated Gramsat project in Orissa, it aims at empowering the people especially the poor and marginalized, by awareness building and access to information and services.
- Acreage estimates and crop inventory is being done during *kharif* and *rabi* season for rice.

- Other crops like banana, chillies, cotton, maize, sugarcane and tobacco are also being inventoried.

In the years to come, precision agriculture may help the Indian farmers to harvest the fruits of frontier technologies without compromising the quality of land and produce. The adoption of such novel technique would trigger a techno-green revolution in India which is the need of the hour. The PF and RS has been found potentially important in agriculture applications.

8.16 KEY REFERENCES AND RESOURCES FOR FURTHER READING

1. Australian centre for precision agriculture (ACPA) 1999. Precision agriculture. An introduction to concepts, analysis and interpretation (1999).
2. Baburao DK, 2004. Precision farming in Indian Agricultural Scenario. Remote sensing and geographic information system. Asian Institute of Technology, Thailand (on line).
3. Biswas C, Subba Rao AVM. GIS and its role in Agriculture. Yojana 2000; 24–25 (June edition).
4. Cambouris AN, Walin MC, Simard RR. In proceedings of the 4th International conference on precision agriculture, St. Paul, MN (Eds. Robert PC, Rust RH, Larson WE, Madison), WI: ASA CSSA-SSSA 1999; p847–858.
5. Das DK. Remote Sensing — Its Application in Agriculture. *Introductory Soil Science*. Kalyani Publishers, India 2007; p144–160.
6. Donald P, Casady W, Shannon K. 1998. Precision Agriculture. Global positioning system (GPS). Published by: University Extension, University of Missouri — Lincoln University.
7. Ghosh SK, Murthy KMD, Ramesh G et al. Green revolution and its impact on Indian Agriculture Employment News 1999; 222:1–2.
8. Goodchild MF. GIS and Geographic Research. In: Pickles J (ed.) Ground Truth. The Social Implications of Geographic Information Systems. Guildford Press, New York 1995; p31–50.
9. Hatfield JL 2000. Precision agriculture and environmental quality. Challenges for research and education: National Workshop, "Precision Agriculture and the environment: Research properties of the nation. USDA's Natural Resource Conservation Service and Agricultural Research Service.
10. Heimlich R. GIS and Agriculture. *Agriculture Outlook April* 1998; 19–23.
11. Hess JR, Svobada JM, Hoskinson RO, *et al*. Site Specific Planting. In: Proceedings of the 4th International Conference on Precision Agriculture (Eds. Robert PC, Rust, larson WE), ASA-CSSA, Madson, W.I. 1999; p653–660.
12. Jat ML, Pal SS, Subba Rao, *et al*., 2003. In: National seminar on developments in soil science, 68th annual convention of the Indian society of soil science, CSAUAT, Kanpur.
13. Jat ML, Pal SS, Subba Rao, *et al*. In: Proceedings of National Conference on Conservation Agriculture: Conserving Resources — Enhancing Productivity, NASC Complex, Pusa New Delhi 2004; p9–10.
14. Jat ML, Sharma SK, Gupta RK, *et al*., 2005. In: Special Publication on Conservation Agriculture, CASA, NASC, New Delhi.
15. Khosla R. Colarado State University Agronomy Newsletter 2001; 21(1):24–26.

16. Mishra A, Raj PC, Balaji D 2008. Operationalization of Precision Farming in India. Department of Civil Engineering, College of Engineering, Guindy, Anna University, Chennai–600025.
17. Mulla DJ. Geostatistics, Remote Sensing and Precision Farming. In precision agriculture: spatial and temporal variability of environmental quality (Eds Lake JV, Bock GR, Goode JA). John Wiley & Sons, New York 1997; p100–119.
18. Narayan LRA. Survey of Indian Agriculture. The Hindu 2000; 193–195.
19. Pal SS, Rao Subba, Jat AVM, *et al.* In proceedings of national symposium on alternate farming system. Enhanced income and employment generation options for small and marginal farmers, PDCSR, Modipuran 2004; p227–28.
20. Racher C 1999. Precision Farming Primer. Published by University of Monnesatto; Directorate of Extension Education.
21. Reddy GP, Anand PSB. Precision Farming and GIS Techniques, 2000; *Yojana* 35–36 (June edition).
22. Schepers JS 2000. Environmental Problems Facing Agriculture. National Workshop, "Precision agriculture and the environment: Research properties of the nation. USDA's Natural Resource Conservation Service and Agricultural Research Service.
23. Sharma SK, Biswas C. Precision Farming and its Relevance in Indian Agriculture. *Indian Journal of Soil Fertility* 2005; 1(4):13–26.
24. Shukla AK, Singh VK, Dwivedi BS, *et al.* SSNM in Hybrid Rice Wheat Cropping System. *Better Crops* 2004; 88(4):18–21.
25. Singh KK, Shekhawat MS. Precision Agriculture and Information Technology. *Farmer and Parliament* 2000; 10–13.
26. Singh KK, Khan M, Shekhawat MS. Agriculture Scenario in India. *Yojana*, p25–28 (June edition).
27. Singh VK, Jat ML, Sharma SK. In: Rice wheat information sheet, rice-wheat consortium for the Indo-Gangetic plains, NASC, Pusa, New Delhi 2005; p5.
28. Thirkawala S, Weer Sink A, Kachaoski G, *et al.* GPS and its Role in Agriculture. *American Journal of Economics*, 1999; 81:914–927.

SOME THOUGHTS ABOUT THE EARTH/SOILS

- Hinduism: "Kshiti, Jal, Pawak, Gagan, Sameera, Panch tatwa yah bana sareer of (Prasnav Upanishad)"
- Buddhism: "One should not even break the branch of a tree that has given one shelter" (Petavatthu 11, 9, 3)
- Islam: "He created the man of clay like potter" (Suhrah Al-Rhman, Verse 14)

  "We made from water everything (Quran 25:54)"

  "Do not overuse water even if you are on a running river" (Prophet Mohanmad)
- Greek: The daughter of Earth goddess "Gaia" named Themis (Goddness of law) and her descendants demeter was the goddess of agriculture and fertility"
- Roman: The Earth goddess (Tellus) was related to the goddess of fertility and harvest (ceres)

CHAPTER

9 Impact of Deforestation on Flora and Soil Attributes — A Critical Review

Forest are lives of million people. —Anonymous

Forests are one of the most valuable ecosystems in the world, containing over 60 per cent of the world's biodiversity. This biodiversity has multiple social and economic values, apart from its intrinsic value, varying from the important ecological functions of forests in terms of soil and watershed protection to the economic value of the numerous products, which can be extracted from the forests. Forests play a crucial role in climate regulation and constitute one of the major carbon sinks on earth, preventing increase in greenhouse effect. Deforestation causes various disturbances in our ecology and need immediate efforts towards sustainable forest development for ecological and social well being of human kind. This article highlights critical review on deforestation impact on flora and soil properties in general and in Himalayan region in particular.

9.1 INTRODUCTION

Forests are playing a vital role in the general economy of the country; they also act as valuable resources for providing subsistence to rural people (Shah, 1984). Further, protective influence of forests on habitat not only immediately under their canopy, but also far considerable distances around has been generally accepted (Jha, 1998). Despite their innumerable benefits — tangible and intangible, man has been clearing forests ruthlessly for quick monitory gains (Lal, 1989). During the year 1980s, deforestation was high as 15 million hectares per year for tropical forests alone and during 1990s it developed highly (Anonymous, 2005). Deforestation permanently destroys the biodiversity that a forest contains and a degraded forest may not be able to support species adapted to the specialized condition (Pimm and Askins, 1995). Deforestation causes remarkable change in vegetation diversity and species composition (Verma *et al.*, 1997). Forests support many species that are only found in a small geographical range, so any disruption can result in loss of species or loss of genetic variability within the species (Falk, 1990) — a subtle loss to biodiversity. Loss of individual species can have variously different effects on the remaining species in an ecosystem (Shekharia, 2004). Deforestation also increases soil erosion causing soil nutrient loss.

Forests in India occupy 7.6 lakh sq. km of the total geographical area (Anonymous, 2001 a). In Jammu and Kashmir province; Kashmir has an area of 8128 sq. km under forest cover against 12066 and 36 sq. km of Jammu and Ladakh, respectively (Anonymous, 2000). Forests in Kashmir, occupying importance second to agriculture, harbor in rich herbaceous flora. But due to deforestation, forest cover has reduced to 51 per cent (for the year 1997–1998) against 59 per cent during the pre-planned era, affecting herbaceous wealth of these ecosystems (Anonymous, 2001b). Dhara catchment (present study area) extending over an area of 21467 hectares is one of the badly deforested areas of the valley.

9.2 IMPACT OF DEFORESTATION ON FLORA

Deforestation not only destroys biodiversity but affects habitat as well. According to Buschbacher (1986) the rate of Amazon deforestation in a global context, the environmental consequences of deforestation and the potential of the cleared land for sustainable production of food and protein, it is concluded that loss of biological and cultural diversity is the most detrimental effect in the area. Arnalds (1987) gave a review of changes in the vegetation since settlement in 874 AD, the nature of disturbances that led to changes in vegetation and landscape, and the distribution and status of the current vegetation in Iceland. Birch (*Betula pubescens*) woodlands were thought to have covered 25 per cent of the country, but now cover only 1 per cent of the land area. About 65 per cent of Iceland was vegetated at the time of settlement, but this cover has reduced to <25 per cent. It is concluded that extensive soil erosion due to deforestation and continued degradation of vegetation by livestock grazing are the most serious environmental problems in Iceland.

Biswas (1988) while studying the rare and threatened taxa in the forest flora of Tehri Garhwal Himalaya and the strategy for their conservation reported that it is estimated that over 25,000 species of plants are on the verge of extinction world over. The grave threats to the rare, threatened and endangered plants are due to many factors, such as destruction of natural habitats due to acute pressure on land and forests by a rapid increase in human population and over collection of species of horticultural importance. He has also given the record of collection and status of 25 taxa facing extinction or are vulnerable and rare in the area. *Castanopsis tribuloides, Clematis connata, Indigofera cassioides and Mahonia jaunsarensis* are the species suspected to have undergone extinction.

Though much of the loss of permanent vegetation is taking place in areas outside the legally classified forests (Lal *et al.*, 1990), however according to World Rainforest Movement (1990) deforestation is causing a loss of biological diversity on an unprecedented scale. Although tropical forests cover only 6 per cent of earth's land surface, they happen to contain between 70 and 90 per cent of all of the world's species. Due to deforestation, we are loosing between 50 and 100 animal and plant species each day. Many of the species now facing the possibility of extinctions are of enormous potential to humans in many areas, especially medicines. Sioli (1991) while studying the foreseeable effects of deforesting Amazonia reported extermination of the richest and most diverse gene stock of the earth by the extinction atleast hundreds of thousands, if not millions, of plants species. This effect due to deforestation means a massive impoverishment of the life of Amazonia and of the life forms of bio productivity. Gliessman *et al.*, (1993) while studying the "Managing diversity in traditional agro-ecosystems of tropical Mexico" reported that it has been estimated that 90 per cent of the forests in the lowland humid tropics of Mexico had disappeared by the 1970s, due to causes ranging from cattle ranching to oil exploitation and during the last decade, 17 plant species have become extinct and another 477 are endangered, representing 17 per cent of the floristic richness of Mexico.

Since habitat requirements of most species are quiet narrow, with the loss of suitable habitats, populations are destroyed eventually leading to extinction of species. It was estimated that the extinction rates based on relation between habitat loss and species loss indicate that regions rich in endemic species dominate the global patterns of extinction. According to the cookie-cutter model proposed by the Pimm *et al.*, when habitat destruction cuts out areas, species also found elsewhere survive as they can recognize but some of the endemic species become locally extinct, the proportion depending on the extent of habitat loss (Pimm *et al.*, 1995).

Vandermeer and Perfecto (1995) reported that the effect of selective logging is the changes that occur in the forest when a gap forms due to the loss of a tree. For tropical rainforests, the hole created by a fallen tree leads to changes that eventually affect the biodiversity of the habitat. The tree that previously occupied the space preserved a dark area of land within the forest. With the canopy of the tree to provide shade gone, light streams in and causes plants that do better in brighter conditions to force out low light plants that existed there previously. The loss of low light plants leads to a decline in biodiversity.

Sinha (1996) stated that through indiscriminate exploitation, destruction of habitats spread of harmful chemicals and introduction of alien species, a number of plants have already disappeared while others await a similar fate, at times, even without our being aware of their existence. Thus, human interference has been more often responsible for the depletion of pant resources and consequent decline of genetic diversity. According to an estimate of the Threatened Plants Committee of the international union for conservation of natural resources (IUCN) about 10 per cent (20,000 to 30,000) of the world's flowering plants are reported to be "dangerously rare" or "under threat". According to Singh and Vishwakarma (1997) Indian flora comprises around 15,000 flowering plants, an estimated 33 per cent of which are endemic and represent 6 per cent of the world's total. As many as 3000 to 4000 plants are endangered.

Sharma (2000) studied endangered species from Rajasthan (India) belonging to 67 taxa of angiosperms, one taxon of *Pteridophyta* pertaining to 63 genera and 33 families. The endangered status of these species is attributed to the destruction of the forest vegetation. *Pterocarpus marsupium,* which was previously common in forest, has become extinct. Due to deforestation the estimated species loss rates from forest fragments were: 49.8 per cent of common species from one hectare fragments, 29.8 per cent from 10 hectare fragments and 13.8 per cent from 100 hectare fragments (Didham *et al.*, 1998). Sinha and Singh (1999) reported that the greatest threats to biological diversity are the destruction of ecosystems, especially in the tropics and the disappearance of habitats in the wake of developmental activities. They also reported that a large number of species are also threatened by over hunting, poaching and illegal trade.

Biological extinction has been a natural phenomenon in geological history. But the rate of extinction was perhaps one species every 1000 years. But man's intervention has speeded up extinction rates all the more. Between 1600 and 1950, the rate of extinction went up to one species every 10 years. Currently it is perhaps one species per year (Agarwal, 2000). Further, Puig (2001) reported an increase in soil erosion and assessment of biodiversity losses from 0.5 to 1 per cent per annum in tropical zones due to deforestation. Brooks *et al.*, (2002) observed that nearly half the world's vascular plant species and one third of the terrestrial vertebrates are endemic to 25 'hotspots' of biodiversity, each of which has atleast 1500 endemic plant species. None of these hotspots have more than one-third of their pristine habitat remaining. Historically, they cover 12 per cent of the land's surface, but today their intact habitat covers only 1.4 per cent of the land. As a result of this habitat loss, we expect many of the hotspot endemics to have either become extinct or because much of the habitat loss is recent to be threatened with extinction. Pullaiah (2002) reported that during the next 20 to 30 years the world could lose more than a million species of plants primarily because of environmental changes due to humans. At 100 species per day, this extinction rate will be more than 1000 times the estimated "normal" rate of the extinction. He also added that about 10 per cent of temperate region plant species and 11 per cent of world's 9000 bird species are at same risk of extinction. In the tropics, the destruction of forests, threatens 1,30,000 species which live nowhere else.

Looming mass extinction of biodiversity in the humid tropics is the major concern for the future. As per reports on local extinctions among a wide range of terrestrial and fresh water taxa from Singapore (540 sq km) in the relation to habitat loss exceeding 95 per cent over 183 years. (Brook *et al.*, 2003) Extinctions were generally fewer, but inferred losses often higher, in vascular plants, plasmids, decapods, amphibians and reptiles (5%–8%). Forest reserve comprising only 0.25 per cent of Singapore's area now harbor over 50 per cent of the residual native biodiversity. Margaret *et al.*, (2003) while studying "Countryside bibliography of neotropical herbaceous and shrubby plants" observed that understory and pasture plots were consistently species poor, while tree-fall gaps and road verges near forest were consistently the most species-rich habitats. In each study area, they found the same proportion of species restricted to forested habitats (45%) and deforested habitats

(37%), and the same proportion of "countryside-habitat generalists" (18%) occurring in both forested and deforested habitats.

According to the NASA Earth Observatory (2003) half of all 5 to 80 million species live in the rainforests. Rainforests only make up seven per cent of Earth's total land area, making these habitats dense with life. Scientists have only named 1.5 million species in detail but yet about 137 species become extinct daily mainly due to habitat destruction. Extrapolations of the observed and inferred local extinction data, using a calibrated species area model, imply that the current unprecedented rate of habitat destruction in Southeast Asia will result in the loss of 13–42 per cent of regional populations over the next century, at least half of which will represent global species extinctions.

In tropical Asia 65 per cent habitat destruction has taken place due to deforestation (Ralapalli and Geeta, 2003). Due to deforestation shade loving species get exposed to sunlight resulting in the disturbance of their microenvironment and thus their population reduces. World Conservation Union (2004) warn that the world's biodiversity is declining at an unprecedented rate. According to the report at least 15 species have gone extinct in the past 20 years and another 12 survive only in captivity. Ghosh *et al.*, (2005) studied the effects of biotic disturbance, such as deforestation on the botanical composition in Goalpara District, Assam, India. A list of herbaceous 'increase' and 'decrease' as well as the pioneer species observed at the study area was presented. The total density in the protected and unprotected areas was 70.3 and 51.6, respectively.

Klink and Machado (2005) reported that the Cerrado was one of the world's biodiversity hotspots and in the last 35 years, more than 50 per cent of its approximately 2 million sq km has been transformed into pasture and agricultural land while studying conservation of the Brazilian Cerrado. The Cerrado has the richest flora among the world's savannas (>7000 species) and high levels of endemism. Deforestation rates have been higher in the Cerrado than in the Amazon rain forest. Only 2.2 per cent of its area is under legal protection. Numerous plant species are threatened with extinction, and an estimated 20 per cent of threatened and endemic species do not occur in protected areas. Soil erosion and the degradation of the diverse Cerrado vegetation formations and the spread of exotic grasses are widespread and major threat to Cerrado. Kumar and Mahindra (2005) reported that every species has its importance in its ecosystem and it can provide new genetic material for improvement. Economically important plants were overexploited to meet the demand of growing population throughout the globe and resulted in drastic decline in the size of their populations. Some species of important plants have already become extinct and there are many facing danger of extinction. Many factors both natural and man-made including deforestation have been responsible for limiting the distribution of species and causing them to become rare or even extinct. Verma *et al.*, (2005) while analyzing the plant diversity in degraded and plantation forests in Kunihar forest division of Himachal Pradesh reported that the number of herb species under plantation forest and degraded forest was 31 m^{-2} and 25 m^{-2}, respectively. On the basis of importance value index (IVI), *Justicia simplex* and *Andropogon* spp. were observed to be the dominant herbs under plantation forest and degraded forest, respectively. Further, observed that index of diversity for herbs was 4.40 in plantation forest and 3.70 in degraded forest. The index of dissimilarity for herbs between plantation forest and degraded forest was high indicating remarkable degree of dissimilarity in herb species.

9.3 IMPACT OF DEFORESTATION ON SOIL PROPERTIES

9.3.1 Soil Reaction

Sambyal and Sharma (1986) reported the pH range of 5.8 to 7.8 in lower Himalayan eroded soils of Himachal Pradesh. On the other hand, a pH range of 3.5 to 5.1 in high altitude forests soil of Sikkim and 6.3 to 8.0 pH of Kashmir valley were reported by Gangopadhayay *et al.*, (1986) and Pal and Deshpande (1987), respectively. According to Sharma and Qather (1989) the pH varied from 4.6 to 5.8 in some outer Himalayan protected and eroded forest soils of Himachal Pradesh, while Verma *et al.*, (1990) reported a pH variance of 5.4 to 6.7 in some forest soils of Kashmir valley and 6.5 to 7.8 in Northwest Himalayan soils (Singh *et al.*, 1991). Das *et al.*, (1992) studied the soils under different aged plantations of *Pinus patula* and *Cryptomeria japonica* in Takdah range area of Darjeeling Forest Division and reported that all the soils were highly acidic. Surface soils under old plantation had a lower pH than those under young ones. The decomposition of litter derived from these species had an acidifying influence on soils and due to more litter fall under old plantations lower pH value was observed.

Minhas *et al.*, (1997) reported a pH variance of 5.0 to 5.8 and 5.3 to 5.8 under forest and cultivated soils of wet temperate zone of Himachal Pradesh. Soil pH varied from 6.7 to 7.5 and 6.3–8.2 in surface and sub-surface layers of temperate zone in Himalayan forest, respectively. In general, pH increases down the profile and this can be attributed to increase in $CaCO_3$ in the sub-surface horizons (Kaushal *et al.*, 1997). Le Bissonnais and Arrouays (1997) reported a pH range of 5.0 to 6.9 in some soils of Southwest France. Whereas, in humid temperate forest soils of Germany, Sweden, and Italy, the pH was observed in the range of 3.0 to 5.6 (Matschonat and Vogt, 1997). Ajaz (2003) while studying the soil characteristics in relation to erodibility in Dhara watershed of Dal catchment reported that the soil pH under forest land use for the surface soils at 10, 20 and 30 per cent slopes ranged from 6.6 to 7.2, 6.2 to 7.2, 6.7 to 7.5 with corresponding average values being 6.8, 6.7 and 7.1. While in sub-surface it ranged from 6.3 to 7.4, 6.6 to 7.5 and 6.5 to 7.4 with mean values being 6.9, 6.9 and 6.9, respectively. The average pH values for the surface and sub-surface, for forestland use soils at different slopes were 6.9 and 6.9, respectively.

Verma *et al.*, (2005) while analyzing plant diversity in degraded and plantation forests in Kunihar Forest Division of Himachal Pradesh observed pH values of

7.79 and 7.24 for plantation forest and degraded forest, respectively. Further, Zargar *et al.*, (2005) reported the impact of degradation on physicochemical characteristics and microbial status of forest soils dominated by Fir and Spruce reported pH in neutral range (6.8–7.6) in all established sites whereas in degraded sites, pH was in alkaline range (9.2).

9.3.2 Organic Carbon

Sachan *et al.*, (1981) while studying some hill and forest soils of Himachal Pradesh found that the forest soils have the bulk of their organic carbon close to the soil surface as against the grassland. Organic carbon ranges between 0.62 to 6.4 per cent in some high altitude forest soils of Sikkim, and sharp fall in the contents of organic carbon with increasing depth (Gangopadhayay *et al.*, 1986). According to Lahiri and Chakravarti (1989) high altitude soils are possessing a high organic carbon content than soils of lower altitude and attributed it to low temperature and heavy rainfall which restricts the microbial population. Mandal *et al.*, (1990) while studying the soils of middle and upper hill forests of the Eastern Himalayas found that soils were rich in organic carbon content particularly in the upper horizon of upper hill forests than in middle hill forest soils. Cerri *et al.*, (1991) while studying nature and behavior of organic matter in soil under natural forest and after deforestation concluded that the quick response of soil organic matter to external influences, such as deforestation makes it a powerful tool to track related soil changes; mineralization and eluviation processes are no longer balanced by litter fall and decay, resulting in the subsequent decrease of soil humus content. Matirns *et al.*, (1991) concluded that the main factor that affects the clear felled soils is probably the removal of the natural vegetation cover, rather than the crop itself. Two facts seem to prevail that there is decrease in soil organic matter and the decrease in soil biological activity which negatively affects aggregate stability.

The organic carbon content of some soils under forests of Kashmir valley vary from 0.3 to 5.4 per cent (Verma *et al.*, 1990); 0.15 to 2.58 per cent in soils of North-Western Himalayas (Singh *et al.*, 1991); 0.68 to 1.48 in mustard growing soils of North Kashmir (Kher and Singh, 1993); 1.0 to 7.7 per cent in few entisols and inceptisols of North-Western Himalayas (Kaistha and Gupta. 1994); 0.12 to 5.37 per cent in forest soils of wet temperate zone of Himachal Pradesh (Minhas *et al.*, 1997). Das *et al.*, (1992) studied the soils under two different types of coniferous species of Takdah area Darjeeling Forest Division and found that the contents of organic carbon was found to be much higher in the upper layers than in the sub soil. Saikh *et al.*, (1998) while studying the soils of Simlipal National Park, India reported that in deciduous forests and evergreen forests, average values of organic carbon are 3.52 per cent and 5.97 per cent, respectively. Ajaz (2003) reported that the organic carbon content under forest land use for the surface soils at 10, 20 and 30 per cent slopes varied from 1.58 to 2.39, 1.68 to 2.59 and 2.03 to 2.66 with mean values of 2.20, 2.32 and 1.95. Whereas in sub-surface soil it varied from 1.38 to 2.14, 1.38 to 2.19 and 1.04 to 2.31 with corresponding average values being 1.65, 1.96 and

1.95 per cent, respectively. The average values for organic carbon content for different slopes surface and sub-surface soils of forest land were 2.31 and 1.86 per cent, respectively.

9.3.3 Available Nitrogen

Gupta (1974) reported that the available nitrogen ranges from 50 to 200 ppm in soils of Ranbir Singh Pora of Jammu and Kashmir. Whereas in alluvial soils, the available nitrogen varied from 36 to 205 ppm with a mean value of 118 ppm (Verma *et al.*, 1978). Mandal *et al.*, (1990) reported that available nitrogen is in the range of 31 to 221 ppm. They further reported that the available nitrogen was highest in surface horizons and regularly decreases with depth. Khanthaliya and Bhat (1991) reported positive correlation between organic carbon and available nitrogen in soils of sub-humid zone which may be due to association of nitrogen with organic matter and absorption of NH_4-N by humus complex in the soils. Verma *et al.*, (2005) reported that soil under plantation forest have better fertility status than that of degraded forests. The available nitrogen found in plantation forest and degraded forest was 222.85 and 114.96 kg ha^{-1}, respectively. Available nitrogen of degraded soils was 71 and 53 per cent less than those of established forest soils as reported by Zargar *et al.*, (2005).

9.3.4 Available Phosphorus

The available phosphorus ranged from 2.83 to 12.20 ppm in some soils of UP (Bhan and Shanker, 1973) and from 3.73 to 49.5 ppm in alluvial tracks of UP as observed by Singh and Pathak (1973). Further, Gupta (1974) reported that available phosphorus ranged from 9.15 to 30.0 ppm in soils of Jammu and Kashmir State. The significant positive relationship of total phosphorus with organic carbon and cation exchange capacity has been reported by Mandal *et al.*, (1990). However, available phosphorus did not show any relation with pH and total phosphorus. Verma *et al.*, (2005) reported that the available phosphorus found in plantation forest and degraded forest was 65.75 and 40.66 kg ha^{-1}, respectively. Zargar *et al.*, (2005) reported that the available phosphorus of degraded soils ranged between 1.3–14 ppm and in established forest soils the available phosphorus ranged between 6.15–31.0 ppm.

9.3.5 Available Potassium

The available potassium content varied from 8.4 to 286.1 ppm in the foot hill soils of Gurdaspur and Hoshiarpur districts of Punjab (Singh and Raman, 1982) 10.5 to 476.3 ppm in soils of Himachal Pradesh (Thakur and Bhandari, 1986) and 28 to 520 ppm in some soils of Haryana (Kumar *et al.*, 1987). Similarly, Mandal *et al.*, (1990) have also reported a range of 47.2 to 289.8 ppm of available K in some forest soils of Eastern Himalayas. Verma *et al.*, (2005) reported that the available potassium in plantation forest soil and degraded forest soil in Kunihar Division of Himachal Pradesh was 63.85 and 49.26 kg ha^{-1}, respectively. On the other hand Zargar *et al.*, (2005) have observed on impact, degradation of physiochemical characteristics and microbial status of forest soils dominated by fir and spruce and also noticed that the available potassium of established forest soils ranged between 90 and 425 ppm whereas degraded forest soils have 30–320 ppm.

9.4 CONCLUSIONS

Deforestaion is an important factor in deciding soil fertility and productivity.

9.5 KEY REFERENCES AND RESOURCES FOR FURTHER READING

1. Agni T, Pandit A, Pant K, *et al*. Analysis of Tree Vegetation in the Tarai-Bhabhar Tract of Kumaun Central Himalaya. *Indian Journal of Forestry*, 2000; 2(3):252–261.
2. Agnihotri Y, Yadav RP, Prasad R *et al*. Effect of Aspect and Physiographic Position on Vegetation Cover in Shivalik Watershed at Relmajra, Punjab. *Indian Journal of Forestry*, 2006; 29(1):9–13.
3. Ahmed G, and Bhuyian MK. Regeneration Status in the Natural Forests of Cox's Bazar Forest Division, Bangladesh. *Annals of Forestry*, 1994 2(2):103–108.
4. Ajaz G S, 2003. Studies on Soil Characteristics in Relation to Erodibility in Dhara Watershed of Dal Catchment. M.Sc. thesis submitted to Sher-e-Kashmir University of Agricultural Sciences and Technology of Kashmir, Shalimar, Srinagar p85–94.
5. Kumar A, Puri S, and Kumar R. Performance of Indigenous and Exotic Tree Species and Their Impact on Soils of Arid Region. *Crop Research Hissar*, 1994; 7(2):292–298.
6. Anonymous. Spruce Regeneration and Soil Science. *Rev For France*, 1953; 5(4):257–268.
7. Anonymous, 1993. Annual Progress Report. Division of forestry, SKUAST-K, Shalimar, Srinagar, Kashmir.
8. Anonymous, 2000-01. Annual Administrative Report, Forest department, Statistics Division, Srinagar, Kashmir p2.
9. Anonymous, 2002–03. Digest of Statistics. Directorate of Economics and Statistics (Planning and Development Department). Government of Jammu and Kashmir. DOS (28):03, p109.
10. Anonymous, 2005. State of Forest Report 2003. Forest Survey of India, Ministry of Environment and Forests, Dehradun, p134.
11. Aquirre JL, Bartolome C, Alverez J, *et al*. Mortality of Beech (*Fagus sylvatica*) Seedling During the Summer Months in the Central Mountain System in Spain. *Studia Oecologica*, 1991; 8:127–138.
12. Assis RL, Ferreira MM, Morais EJ, *et al*. Behaviour of Soil Moisture and Water Storage of *Eucalyptus urophylla* Plantations at Different Spacings Compared with the Cerrado Vegetation at Bocaiuva (MG). *Ciencia-e-Agrotecnologia*, 1998; 22(1):79–86.
13. Banerjee SK, Kashyap MK, and Sahai A. Vegetation Status, Nutritional Characteristics and Biological Activity of Manganese Mine Over Burden Spoils in Relation to Age. *Indian Forester*, 2000; 126(7):742–748.
14. Banerjee SK, Nath S, and Banerjee SP. Characterization of Soils Under Different Vegetations in the Tarai region of Kurseong forest division, West Bengal. *Journal of the Indian Society of Soil Science*, 1986; 34(1):343–349.

15. Bean W, and Riotte L 1981. Companion Planting for Successful Gardening. Garden Way, Vermont, USA.
16. Behari B, Aggarwal R, Singh AK, *et al*. Vegetation Development in a Degraded Area Under Bamboo Based Agroforestry System. *Indian Forester*, 2000; 126(7):710–720.
17. Bentham C, and Hooker DJ 1897. Flora of British India. L. Reeve and Co. Ltd. The Oast house, Brook, NR, Ashford, Kent, England.
18. Bhan C, and Shanker H. Studies on Forms and Contents of Soil Phosphorus and Their Inter-relationship with Some Physico-Chemical Characteristics of Selected Soils of UP. *Journal of the Indian Society of Soil Science*, 1973; 39(4):781–782.
19. Bhandari BS, Mehta JP, Nautiyal BP, *et al*. Structure of a Chir Pine Community Along in Altitudinal Gradient in Garhwal Himalaya. *International Journal of Ecology and Environmental Sciences*, 1997; 23(1):67–74.
20. Bhuyan P, Khan ML, and Tripathi RS. Regeneration Status and Population Structure of Rudraksh (*Elaeocarpus ganitrus* Roxb.) in Relation to Cultural Disturbances in Tropical Wet Evergreen Forest of Arunachal Pradesh. *Current Science*, 2002; 83(11):1391–1394.
21. Black CA 1965. Methods of Soil Analysis. American Society of Agronomy. University Wisconsin, USA.
22. Blamey M, and Wilson G 1989. Trees and Shrubs Hardy in the British Isles. London, ISB No. 7195-1790–7.
23. Chandra KK, Bhandari PS, Khosla H, *et al*. Comparative Study on Regeneration Status, pH and Organic Matter Content in Protected Mixed Forest and Unprotected Forest. *Indian Journal of Forestry*, 2001; 24(4):414–418.
24. Curtis JT, and McIntosh RP. The Inter-relations of Certain Analysis and Synthetic Phyto-sociological Characters. *Ecology*, 1950; 31:434–455.
25. Dar GH, and Christensen KI. Habitat Diversity and Zonality of Vegetation in Sind Calley, Kashmir Himalaya. *Nature and Biosphere*, 1999; 4(1&2):49–71.
26. Das PK, Nath S, and Banerjee SK. Soil Characteristics Under Introduced Pinus Patula in Darjeeling Himalayan region. *Indian Agriculture*, 1990; 34(11):25–33.
27. Das PK, Nath S, Sahai RMN, *et al*. Characteristics of Soil Under Different Aged Plantations of *Pinus patula* and *Cryptomeria japonica* in Eastern Himalayas. *Indian Journal of Forestry*, 1992; 15(2):111-115.
28. Dhar O, Jha MN, and Suri S 2001. Soils of the Shankaracharya Hill Forest Research Range Natural Resources of Jammu and Kashmir. A case study published by Altaf Ahmad Bazaz for valley Book House, Srinagar, (J & K) p43–63.
29. Duchok R, Kent K, Khumbongmayum AD, *et al*. Population Structure and Regeneration Status of Medicinal Tree (*Illicium griffithii*) in Relation to Disturbance Gradients in Temperate Broad Leaved Forests of Arunachal Pradesh. *Current Science*, 2005; 89(4):673–676.
30. Dwivedi AP 1993. A Text Book of Silviculture. International book distributors, Dehradun, p286.
31. Gangopadhayay SK, Debnath NC and Banerjee SK. Characteristics of Some High Altitude Soils of Sikkim Forest Division. *Journal of the Indian Society of Soil Science*, 1986; 34(4): 830–838.
32. *Ghosh RC 1977. Handbook on Afforestation Techniques. Controller of Publications, Delhi.

33. Gorden DT 1970. Natural Regeneration of White and Red Fir US forest research paper paof 5th west-for Dange Expt Sta No. PSW 58:32.
34. Guar YD and Satyanarayan NY. Phytosociological Studies of the Monsoon Vegetation of Rock Habitat. *Indian Forester*, 1967; 93(12):806–814.
35. Gupta HS and Tripathy KK. Study on Natural Regeneration in Forests of Garhwa (Bihar). *Indian Journal of Forestry*, 1997; 20(4):324–327.
36. Gupta RD. Studies on Certain Physico-Chemical and Microbiological Properties of Ranbir Singh Pora soils of Jammu and Kashmir. *Technology*, 1974; 11(2 & 3):227–230.
37. Hajra PK and Rao RS 1987. Distribution of Vegetation Types in North-West Himalaya with Brief Remarks on Phytogeography and Floral Resources Conservation. Proceedings of Indian Academy of Science (Plant Science) 100(4):262–277.
38. Hokar VM and Totey NG. Regeneration Status of Navegaon National Park (Maharashtra). *Indian Journal of Forestry*, 1999; 22(3):203–209.
39. Horkar VM and Totey NG. Floristic Diversity and Soil Studies in Navegaon National Park (Maharashtra). *Indian Journal of Forestry*, 2001; 24(4):442–447.
40. Jackson ML 1957. Soil Chemical Analysis, New Delhi. Prentice Hall of India, Pvt. Ltd. p281.
41. Jha AK and Singh JS. Spoil Characteristics and Vegetation Development of an Age Series of Mine Spoils in a Dry Tropical Environment. *Vegetatio*, 1991; 97:63–76.
42. Jha MN, Gupta MK, Dimri BM, *et al.* Moisture Distribution Pattern in the Soil Under Different Tree Plantations. *Indian Forester*, 2001; 127(4):443–449.
43. Kaistha BP and Gupta RD. Morphology and Characteristics of a Few Entisols and Inceptisols of North Western Himalayan region. *Journal of the Indian Society of Soil Science*, 1994; 42(1):100–104.
44. Kale VB and Chevan AS. Physio-Chemical Characteristics of Kumbhave-5. Soil series of Konkan of Maharashtra State. *PKV Res J*, 1996; 20(2):136–140.
45. Kapoor KS, Nath S, Banerjee SP, *et al.* Forest Vegetation of Darjeeling Himalayas — An Analysis. *Van Vigyan*, 1989; 27(1):24–33.
46. Kaul MK 1971. Flowering Plants of Shankaracharya Hill, Srinagar, Kashmir. Bulletin of Botanical Survey of India 13(3&4):236–243.
47. Kaushal R, Bhandari AR, Sharma JC, *et al.* Soil Fertility Status Under Natural Deodar (Cedrus deodara) Forest Ecosystem of North-Western Himalayas. *Indian Journal of Forestry*, 1997; 20(2):105–111.
48. Kent M and Coker P 1992. Vegetation Description and Analysis. Belhaven Press, London. p42.
49. Khan FAS 2007. Status of Vegetation in Shankaracharya Reserved Forest of Kashmir. M.Sc. thesis Submitted to Sher-e-Kashmir University of Agriculture Sciences and Technology of Kashmir Shalimar, Srinagar, Kashmir.
50. Khanthaliya PC and Bhat PL. Relation between Organic Carbon and Available Nutrients in Some Soils of Sub-Humid Zone. *Journal of the Indian Society of Soil Science*, 1991; 39(4):781–782.
51. Kher D and Singh N. Different Forms of Sulphur in Mustard Growing Soils of North Kashmir. *Journal of the Indian Society of Soil Science*, 1993; 41(1):164–165.

52. Kumar BM. Regeneration Status of Some Important Forest Formations in the Western Ghats of Kerala India. *Annals of Forestry*, 1997; 5(2):155–164.
53. Kumar R, Mahandra S, Ruhal DS, *et al*. Forms of Potassium in Some Soils of Central Haryana. *Punjab Agriculture University Journal of Research*, 1987; 17(4):356–363.
54. Kunhikannan C, Verma RK, Khatri PK, *et al*. Ground Flora, Soil Microflora and Fauna Diversity Under Plantation Ecosystem on Bhata land of Bilaspur (MP). *Environment and Ecology*, 1998; 16(3):539–548.
55. Lahiri T and Chakravati SK. Characteristics of Some Soils of Sikkim at Various Altitudes. *Journal of the Indian Society of Soil Science*, 1989, 37(3):451–454.
56. Lall J. Vegetation Structure and Regeneration Studies on Two Adjacent Protected and Unprotected Tropical Forest Sites in Central India. *Indian Forester*, 1990; 116(3):194–201.
57. Le Bissonnais Y and Arrouays D. Aggregate Stability and Assessment of Soil Crustability and Erodibility. *European Journal of Soil Science*, 1997; 48:39–48.
58. Luna RK 2005. Plantation Trees. International Book Distributors, Dehradun.
59. Mandal AK, Nath S, Gupta SK, *et al*. Characteristics and Nutritional Status of Soils of Middle Hill and Upper Hill Forest of the Eastern Himalayas. *Journal of the Indian Society of Soil Science*, 1990; 38(1):100–106.
60. Maschonot G and Vogt R. Effect of Changes in pH Ionic Strength and Sulphatic Concentration on the CEC of Temperate Acid Forest Soils. *European Journal of Soil Science*, 1997; 48:163–171.
61. Merwin HD and Peech M. Exchangeability of Soil Potassium in Sand, Silt and Clay Fraction as Influenced by the Nature of Complementary Exchangeable Cations. *Soil Science Society*, 1951; 15:125–128.
62. Minhas RS and Bora NL. Distribution of Organic Carbon and the Forms of Nitrogen in a Topographic Sequence of Soils of Himachal Pradesh. *Journal of the Indian Society of Soil Science*, 1982; 30(2):135–139.
63. Minhas RS, Minhas H and Verma SD. Soil Characterization in Relation to Forest Vegetation in the Wet Temperate Zone of Himachal Pradesh. *Journal of the Indian Society of Soil Science*, 1997; 45(1):146–151.
64. Mishra R 1968. Ecology Work Book. Oxford and IBH Publishing Co., New Delhi. P244.
65. Misra A, Sharma CM, Sharma SD and Baduni NP. Effect of Aspect on the Structure of Vegetation Community of Moist Bhabhar and Tarai *Shorea robusta* forest in central Himalaya. *Indian Forester*, 2000; 116(6):634–641.
66. Muthoo MK and Wali. Deodar belt of Kashmir Lolab valley. *Indian Forester*, 1963; 116(2):938–941.
67. Naithani HB, Pal RC and Srivastava RK. Vegetation Analysis of the Tirumala Hills Forest, Andra Pradesh. *Indian Forester*, 2006; 132(2):1110–1130.
68. Olsen SR, Cole CV, Watanable FS *et al*. Estimation of Available Phosphorus in Soils by Extraction with Sodium Bicarbonate. In: Methods of Soil Analysis (Ed. Black, C.A.), Madison 1954 and 1965, p1044–1046.
69. Pal DK and Deshpande SB. Parent Material, Mineralogy and Genesis of Two Bench Mark Soils of Kashmir Valley. *Journal of the Indian Society of Soil Science*, 1987; 35(4):690–698.

70. Pande PK. Quantitative Vegetation Analysis as per Aspect and Altitude and Regeneration Behaviour of Tree Species in Garhwal Himalayan Forest. *Annals of Forestry*, 2001; 9(1): 39–52.

71. Pande PK. Regeneration Behaviour of Important Tree Species in Relation to Disturbance in Joint Forest Management Adopted Village — Forests in Satpura Plateau, Madhya Pradesh, India. *Indian Forester*, 2006; 132(1):91–104.

72. Pande PK and Bisht APS. Regeneration Behaviour of Some Forest Ecosystem. *Journal of Tropical Forestry*, 1988; 4(1):74–78.

73. Pande PK, Bisht APS and Sharma SK. Comparative Vegetation Analysis of Some Plantation Ecosystems. *Indian Forester*, 1988; 113(9):379–387.

74. Pande PK, Negi JDS and Sharma SC 1996. Plant Species Diversity and Vegetation Analysis in Moist Temperate Himalayan Forest. Abstracted in first Indian Ecological Congress, New Delhi 27–31 Dec p51.

75. Pande PK, Negi JDS and Sharma SC. Species Diversity, Turn-Over and Resource Apportionment Among Various Plant Species in a Western Himalayan Forest. *Indian Forester*, 2000; 126(7):727–741.

76. Pande PK, Negi JDS and Sharma SC. Plant Species Diversity and Vegetation Analysis in Moist Temperate Himalayan Forest. *Indian Journal of Forestry*, 2001; 24(4):456–470.

77. Pandey JC. Vegetation Analysis in a Mixed Oak Conifer Forest of Central Himalaya. *Indian Journal of Forestry*, 2003; 26(1):60–74.

78. Pathak MC, Bargali SS and Rawat YS. Analysis of Woody Vegetation in High Elevation Oak Forest of Central Himalaya. *Indian Forester*, 1993; 117(20):722–729.

79. Paul TM 2001. Studies on Woody Ornamental Plants for Landscape Use. Ph.D. thesis submitted to Sher-e-Kashmir University of Agricultural Sciences and Technology, Srinagar p20–31.

80. Prakash R 1998. Plantation and Nursery Techniques of Forest Trees. International Book Distributors, Dehradun.

81. Qaiser KN, Mishra VK and Joshi NK. Factors Affecting Natural Regeneration of Ban Oak (*Quercus leucotrichophora* A. Camus ex. Bahadur): 1. seed fall, infestation, losses and regeneration. *Indian Forester*, 1993; 119(12):986–993.

82. Rajwar GS. The Forest Vegetation on the East and West Facing Slopes of Bhagirathi Valley near Uttar Kachi, Garhwal Himalaya. *Advances in Forestry Research in India*, 1982; 1:199–306.

83. Rajwar GS, Manoj D and Pramod K. Regeneration Status of an Oak Forest of Garhwal Himalaya. *Indian Forester*, 1999; 125(6):623–629.

84. Ralhan PK, Saxena AK and Singh JS 1982. Analysis of Forest Vegetation Around Nainital in Kumaun Himalaya. Proceeding of Indian National Sciences Academy 48:121–137.

85. Rau MA 1963. The Vegetation Around Jamnotri in Tehri-Garhwal, UP, Bulletin of Botanical Survey of India 5(3):277–280.

86. Rikhari HC, Tewari JC, Rana BS, *et al*. Woody Vegetation Along an Altitudinal Gradient in Garhwal. *Indian Journal of Forestry*, 1991; 13:322–338.

87. Sachan RS, Sharma RB and Chibber RK. Nature and Status of Organic Matter of Some Hills and Forest Soils of Himachal Pradesh. *Indian Journal of Forestry*, 1981; 4(4):249–252.

88. Sambyal JS and Sharma PD. Characterization of Some Lower Himalayan Eroded Forest Soils. *Journal of the Indian Society of Soil Science*, 1986; 34(1):142–151.

89. Saxena AK and Singh JS. A Phytosociological Analysis for Forest Communities of a Part of Kumaun Himalaya. *Vegetatio*, 1982; 50:3–22.

90. Sharma BM and Jamwal PS 1988. Flora of Upper Lidder Valley of Kashmir. Himalaya vol. 1 Scientific publication, Jodhpur, India p269.

91. Sharma BM and Raj A. Status of Natural Regeneration of *Juniperus macropoda* Boiss. in Ladakh, the cold arid region of western trans-Himalaya. *Indian Journal of Forestry*, 2004; 27(3):237–240.

92. Sharma PD and Qaher AQ. Characterization of Some Outer Himalayan Protected and Eroded Forest Soils. *Journal of the Indian Society of Soil Science*, 1989; 37(1):113–120.

93. Singh A 1991. Ecology of Vegetation Complex of the Corbett National Park, UP Ph.D. Thesis, Kumaun University, Nainital.

94. Singh AK and Singh B. Effect of Parent Rock on Physico-Chemical Properties of the Soils Developed Under Similar Conditions of Climate, Topography and Forest Covers. *Indian Journal of Forestry*, 1991; 14(3):196–201.

95. Singh AK and Totey NG. Physico-Chemical Properties of Bhata Soils of Rajpur (MP) as Affected by Plantation of Different Tree Species. *Journal of Tropical Forestry*, 1985; 1(1):61–69.

96. Singh B and Raman SS. Distribution of Nutrients in Soil Profiles Under Pines in North East Himalayan. *Journal of the Indian Society of Soil Science*, 1982; 30(2):320–325.

97. Singh G and Kachroo P 1976. Forest Floral of Srinagar and Plants of Neighbourhood. Bishen Singh, Mahendra Pal Singh, Dehradun, India, p278.

98. Singh J and Mahajan AG. Natural Regeneration Status in Melghat Forest an Overview. *Indian Forester*, 1998; 124(4):187–192.

99. Singh J, Gupta GN and Prasad KG. Soil Vegetation Relationship Studies in Some Selected Tree Species of Mudumalai Forest Division. *Indian Forester*, 1988; 114(7):390–398.

100. Singh K, Bhandari AR and Tomar KP. Morphology, Genesis and Classification of Some Soils of North-West Himalayas. *Journal of the Indian Society of Soil Science*, 1991; 39(1):139–146.

101. Singh R, Monaba and Sood VK. A Phyto-Sociological Analysis of Shrub Vegetation Under Diffeent Forest Communities Around Shimla (HP). *Indian Journal of Ecology*, 1991; 8(1):9–16.

102. Singh RP and Gupta MK. Soil and Vegetation Study of Lahaul and Spiti cold Desert of Western Himalayas. *Indian Forester*, 1990; 116(10):785–790.

103. Singh RP and Gupta MK. Vegetation Survey and Ecological Studies Under Silver Fir and Spruce Forest of Himachal Pradesh. *Indian Forester*, 1992; 118(7):361–465.

104. Singh RS and Pathak AN. Forms of Phosphorus in Relation to Physico-Chemical Properties in Bangan Soils of districts Deonia Gonakehpur in alluvial tracts of UP *Journal of the Indian Society of Soil Science*, 1973; 121(4):407–411.

105. Singh RV 1982. Fodder Trees of India. Oxford and IBH Publishing Co., New Delhi.

106. Singh SK. Vegetation Structure Under North and South Aspects in the Temperate Zone of Tirthan Valley, Western Himalaya. *Indian Journal of Forestry*, 1998; 21(3):217–223.

107. Singh SP, Adhikari BS and Zobel DB. Biomass, Productivity, Leaf Longevity and Forest Structure in the central Himalaya. *Ecological Monograph*, 1994; 64(4):401–421.

108. Smiet AC. Forest Ecology on Java : Human Impact and Vegetation of Montane Forest. *Journal of Tropical Ecology*, 1992; 8:129–152.

109. Sood VK and Bhatia M. Population Structure and Regeneration Status of Tree Species in Forest Around Shimla, Himachal Pradesh. *Van Vigyan*, 1991; 29(4):223–229.

110. Subbiah BW and Asija GL. A Rapid Procedure for the Estimation of Available Nitrogen in Soil. *Current Science*, 1956; 25(9):259–260.

111. Tewari A. Tree Layer Analysis of Three Major Forming Species of Kumaun Central Himalaya. *Journal of Eco-Biology*, 1998; 11(1):23–28.

112. Thakur NR and Bhandari AR. Available Nutrient Element Status of Soils of Temperate Vegetable Seeds Producing Valley of Himachal Pradesh. *Journal of the Indian Society of Soil Science*, 1986; 34(3):428–430.

113. Tripathi KP, Mehrotra S and Pushpangadan P. Vegetation Characteristics of Tropical Forests of Andaman Islands. *Indian Forester*, 2006; 132(2):165–179.

114. Troup RS 1921. Silviculture of Indian Trees. vol. III, Clarendron Press. Oxford.

115. Varghese AO and Menon ARR. Vegetation Characteristics of Southern Secondary Moist Mixed Deciduous Forests of Agasthyamalai Region Kerela. *Indian Journal of Forestry*, 1998; 21(4):337–344.

116. Verma CP, Tripathi BP and Sharma DP. Organic Carbon as Index to Assess the Nitrogen Status of the Soils. *Journal of the Indian Society of Soil Science*, 1978; 26(1):138–139.

117. Verma KS, Shyampura RL and Jain SP. Characterization of Soils Under Forests of Kashmir valley. *Journal of the Indian Society of Soil Science*, 1990; 38(1):107–115.

118. Verma RK, Kapoor KS, Rawat RS, *et al*. Analysis of Plant Diversity in Degraded and Plantation Forests in Kunihar Forest Division of Himachal Pradesh. *Indian Journal of Forestry*, 2005; 28(1):11–16.

119. Verma RK, Kapoor KS, Subramani SP *et al*. Evaluation of Plant Diversity and Soil Quality Under Plantations Raised in Surface Mined Areas. *Indian Journal of Forestry*, 2004; 27(2):227–233.

120. Wadia DN 1968. The Himalayan mountains: Its Origin and Geographical Relations. Mountains and Rivers of India 21st International Geographical Congress, New Delhi, India.

121. Walkley A and Black IA 1934. An Examination of the Degtjareff Method for Determining Organic Matter and Proposed Modification of the Chromic Acid Titration Method. Soil Science 34:29–38.

122. Zargar MY *et al*. Impact of Degradation on Physico-Chemical Characteristics and Microbial Status of Forest Soils Dominated by Fir and Spruce. *Asian Journal of Microbiology, Biotechnology and Environmental Sciences*, 2005; 7(1):45–78.

Save forests and reduce risk of global warming. Celebrates 5th June as World Environment Day to create awareness about environment issues. —Subhash Chand

CHAPTER

10 Miscellaneous Topics of Interest

Science without religion is lame, religion without science is blind.

— Albert Einstein

10.1 RESOURCE MANAGEMENT TECHNOLOGIES FOR RICE-WHEAT PRODUCTION SYSTEM

Rice and wheat grown sequentially in an annual rotation constitute a rice-wheat cropping system. In annual cycle suitable thermal conditions for both rice and wheat exist in warm-temperate and subtropical areas and high altitudes in the tropics. Total contribution of rice-wheat cropping system in total cereal production is 85 per cent. More than 10.5 million hectare area is occupied by this system in India. It is a dominant cropping system in the Indo-Gangetic and non-Indo Gangetic plains in India. The following resources management technologies have been adopted for wider applicability in Indo-Gangetic plains and other rice-wheat growing lands.

- The best option for effective use of resources and increasing productivity is via adopting a conservation agriculture approach. Resource conserving technologies (RCTs) have become popular in many cropping systems as a response to increasingly limited agricultural production resources and in view of projected demand for agricultural products. Mostly RCTs are aimed at two most crucial natural resources, i.e. water and soil; however, they also affect the efficiency of other production resources and inputs such as labor, farm power and fertilizer.
- Minimum/zero tillage (ZT) is an innovation that offers both conservation of water and energy resources and results in better crop yields by way of reducing mineralization of soil organic matter. There is a significant adoption of zero tillage in rice–wheat system. Precision land leveling, a highly efficient RCT, is equally suitable for dry and wet land leveling which has shown significant saving of irrigation water, efficient use of inputs and higher yields.
- Laser land-leveling is really laser-controlled land-leveling. Although in agriculture most of the time we are talking about putting a smooth surface with a specific slope on the ground. Laser land-leveling is the actual surface finish, can be controlled to very tight tolerances. Laser is a device that produces a very

concentrated beam of light. Where a common household light bulb produces diffuse light, a laser produces a single, very thin, high energy beam. Instruments can be made that will respond to the energy of a laser beam. The graphical representation of laser land leveling and actual land leveling feature is given in Fig. 10.1. With this the unevenness of fields is reduced to about ±2 cm, resulting in better water application and distribution efficiency, better fertilizer efficiency and reduced weed pressure. Water savings of upto 50% have been reported in wheat and 68% in rice.

Fig. 10.1 Graphical view of laser land leveling

- The advantages of bed planting are saving of irrigation water due to reduced evaporation surface. Water savings to the extent of 42% for transplanted rice and 26% for wheat in bed planting, as compared to flat surfaces, increase yields upto 6.2% for rice and 6.4% for wheat. Direct seeding compared with transplanted rice saves water and labour cost as there is no puddling and transplanting. Growing period from seed to seed is reduced by 10 days; and yields and water efficiency of the following rotation crops, other than rice, are increased.
- Crop residue management is an essential component in rice-wheat system towards conservation agriculture and ensures adequate retention of soil cover. System of rice intensification (SRI) increases rice production and raises the productivity of land, labor, water and capital. Besides saving inputs, it also saves one-third of water applied in conventional system. SRI, which involves planting of single seeding at one hill rather than planting 3 or 4 seedlings per hill, encourages greater root and canopy growth.
- The soil test based recommendations of nutrients coupled with organics and biofertilizers proved beneficial for yield as well as soil health. Dhaencha *(Sesbania aculeate)* and *Sesbania rostrate* are good sources of green manures that can be grown in rice-wheat system. Green manure crops are planted during lag phuse of 60–75 days between harvest of wheat crop and transplanting of rice after 10 July[1].
- During last few years the production as well as productivity of rice-wheat has become static despite the adoption of modern technologies. The farmers need to utilize all the resource conservation technologies for efficient resource or in-

put management for productivity realization. Rice and wheat are the most stable food crops in all the states and need urgent attentions for food security and sustainable development of society. The specialized training needs to develop with linkages of Advanced Center of Agriculture Science and krishi vigyan kendra (KVK) known as modern farm and crop science center at each district levels. The comparative higher productivity and stability of rice-wheat cropping system provides employment and income opportunities to rural masses. Resource management technologies are imperative for adoption and employment for better farmers and public lives in country.

Key References and resources for Further Reading

1. Chand S, Integrated Nutrient Management for Sustaining Crop Productivity and Soil Health. International Book Publishing Company, Lucknow, India, 2008. p1–112.
2. Mishra B. Present Status of Rice and Wheat Research in India. In souvenir of 9th Agriculture Science Congress Organized by NAAS and SKUAST — Kashmir, Srinagar, 22–24 June, 2009. p41–44.

10.2 BENEFICIAL EFFECTS OF SYSTEMS OF RICE INTENSIFICATION (SRI) TECHNOLOGY

Nothing affects the voluntary uptake of conservation agriculture more than crop yields and the net returns earned by practicising farmers, regardless of government or other incentives— Baker 2001

"The system of rice intensification (SRI) is a method of agronomic management of rice cultivation for increasing the yield of rice per unit area per unit time with special and mechanical arrangement, reduced seed and water requirement and modified soil (field) ecosystem." Although SRI is best explained operationally in terms of making certain changes in conventional rice-growing practices, as listed below, it is not best defined in terms of practices. SRI is better understood by focusing on its objectives than on its means. *SRI is a strategy of irrigated rice production, adapted to local conditions, that alters plant, soil, water and nutrient management practices* (the means) with the purpose (the end) of: (a) inducing larger, better-functioning *root systems*, and (b) more abundant, diverse and active communities of *soil biota* that live in association with those root systems. These organisms include both flora and fauna, from scales that are infinitesimally small to visible scales[5]. Indeed, some SRI practices were used in the development of what is called the 'new plant type' that is the use of 14-day-old seedlings, planted singly, and widely spaced, 25 × 25 cm. SRI combines all of these practices, plus it proposes active soil aeration with a rotary weeder[4]. This improved method of rice cultivation was developed in 1983 by the French Jesuit Father Henri de Laulanie in Madagascar and has now spread to many parts of the world. In 1999 introduced in China and Indonesia. In 2002 the International conference on SRI was held in China (15 countries participated including India). 2002 onwards —trials and experiments begun throughout the world. In India pioneer work was initiated at Tamil Nadu Agricultural University

(TNAU) Coimbatore, through the communication of Dr HFM Ten Berge of PRI (plant research international) in the Netherland. Systemic experiments started after 2002 Sanya conference (China). Presently sizeable farmers are practicing SRI in Tamil Nadu, Andhra Pradesh, West Bengal, Punjab and Odisha and also initiated in Gujarat, Uttar Pradesh, Maharastra, Jammu and Kasmir, Bihar, Madhya Pradesh, etc.

- There is a notion that higher yields in rice come with high investments on seed, irrigation, high doses of fertilizers and pesticides. Contrary to this popular view, SRI method of cultivation produces higher yields with less seed and less water. SRI emphasizes on the need to shift from chemical fertilizers to organic manures. Increased soil aeration and organic matter helps in improving soil biology and thus helps in better nutrient availability.
- Pest incidence also reduces due to increased spacing, drastically reducing the need for pesticides. SRI is showing promising results in all rice varieties either local or improved.
- The success of SRI is based on the synergetic development of both the tillers and roots. With a more vigorous root growth, plants can become fuller and taller, and get better access to the nutrients and water they need to produce tillers and seeds. With more growth above ground to carry out photosynthesis, more energy is available for root growth and the stronger the plants, the more resistant they are to attacks by pests and diseases. If two, three or more plants are transplanted together in a clump, competition among their roots limits tillering to 5 per plant at the most. The close planting common in traditional rice cultivation could be considered anti-tillering rice cultivation. To enhance the development of roots and tillers and minimize competition between plants, seedlings are planted one by one in SRI.
- In SRI, spacing between row to row and plant to plant 25 × 25 cm follows a square pattern. In this way a considerable amount of seed can be saved. In SRI, 5 to 8 kg of seed is sufficient for one hectare of transplanted rice, whereas in traditional systems it is quite normal to use 40 to 45 kg/ha in tropical and subtropical areas and 60–80 in temperate regions.
- For centuries rice farmers have kept their paddy fields inundated when their rice is growing. In this way they suppress weeds and reduce the amount of labor needed. This lead farmers and scientists too believe that rice plants benefit from being continuously flooded. However, rice is not an aquatic plant, and although it can survive with its roots submerged it does not really thrive. During its reproductive phase, when plants go through flowering, panicle initiation, grain filling and maturation, maintaining 1 to 2 cm. of water on rice fields has a beneficial effect. But during the preceding growth phase, rice plants grow better in unsaturated soil. The reasons are simple. When there is no standing water and there is air in the soil, the roots can acquire oxygen much more easily through the aerenchyma (air pockets) in the root cells. Lack of oxygen in

the root zone leads to soil acidification that causes the destruction of parilchyma and hampers nutrient up take, assimilation and plant growth. The nitrogen cycle in the soil is disturbed as well, and all kinds of toxicity will develop. Scientists from IRRI have identified the problems caused by anaerobic decomposition in continuously irrigated rice systems as one of the main causes of yield decline. Alternative wetting and drying of the field modifies the growing environment of rice: improves soil structure, gets more oxygen into the root zone, and enhances active soil life. As the soil dries air replaces water and when it rains or irrigation is applied this air is pushed downwards. Periodic water stress and the availability of oxygen facilitate root growth, and the volume of soil penetrated by the roots increases.

- Early weeding in crop is always important for a good return. In rice, where traditional methods are used, hand weeding is usually done one and a half months after transplanting. This is far too late for two important reasons. Not only are weeds replacing half the expected harvest by this time, but farmers also loose the opportunity to bring oxygen into their soil. Aeration of soil by weeding may be even more important in rice cultivation than the removal of weeds. With SRI, simple mechanical push-weeders like kono-weeder, rotary-weeder, etc. are used and these churn up the soil.
- Many of the possible reasons for SRI results pertain to the way that the rice plants are handled differently. Wider spacing, for example, has the effect of achieving *'the edge effect'* throughout the whole field. When plants are more exposed to solar radiation and to circulating air, it is known that this contributes to greater growth and productivity seen in plants growing on the edges of fields. While the edge (or border) effect should be avoided when trying to measure, i.e. estimate, yield, it should be something sought-after through agronomic practice.

The SRI is a resource conserving or rather say input saving technology, having several benefits over traditional planting method of rice cultivation, however adoption in village level is appreciated. It is labor intensive and needs skill during transplanting. Besides several benefits, it helps in restoration of soil health for sustainable crop production system for future agriculture. Such technologies improve socio-economic conditions and rural development in rice growing communities in developing countries for food security.

Key References and resoures for Further Reading

1. WWF–ICRISAT Dialogue Project. 2006. SRI— System of Rice Intensification An Emerging Alternative. Published by Watershed Support Services and Activities Network and Centre for Sustainable Agriculture 12-13-445, Street No.1, Tarnaka, Secunderabad–500 017, Andhra Pradesh, INDIA, 2006.
2. www.cropscience.org.au/icsc2004/symposia/2/4/1869_horiet.htm excess on 15/06/2009

3. Khush GS, Prospects and Approaches to Increasing the Genetic Yield Potential of Rice. In: Rice Research in Asia: Progress and Priorities, eds. Evenson RE, Herdt RW, and Hossain M. Wallingford, UK: CAB International, 57–71. 1996.
4. Singh Lal, System of Rice Intensification for Sustainable Production. In Proc. of Training Course on Crop Diversification for Sustainable Crop Production. Organized at center for advanced studies in agronomy G BPUA&T, Pantnagar. November, 8–28, 2006. p242–246.
5. Randriamiharisoa, Robert, Joeli Barison et al. Soil Biological Contributions to the System of Rice Intensification. In: Biological Approaches to Sustainable Soil Systems, eds. Uphoff N, Ball A, Fernandes ECM, et al. CRC Press, Boca Raton, FL, 409–424. 2006.

10.3 SALINITY IN VERTISOLS

Salinity related land degradation is becoming a serious challenge for food and nutritional security in the developing world. Order vertisols has problem of salinity throughout the country. The vertisols and associates cover nearly 257 million hectares (mha) of the earth's surface out of which about 72 mha occur in India[1]. This shows that nearly 22% of total geographical area of the country is occupied by vertisols. In the central region of India known as the Deccan Plateau, the soils are derived from weathered basalts mixed to some extent with detritus from other rocks. In other areas, particularly in the south, the soils are also derived from basic metamorphic rocks and calcareous clays. Similarly, in the western region, these are derived from marine alluvium that account for nearly 19.6 mha. Out of this about 1.12 mha area affected by salinity and water logging problems[4]. These soils are generally deep to very deep heavy textured with clay content varying from 40%–70%. Further, these are also low in organic carbon content, high in cation exchange capacity, slight to moderate in soil reaction and are generally calcareous in nature. Vertisols, when kept fallow during Kharif season are exposed to soil erosion hazards. Because of their inherent physico-chemical characteristics such as poor hydraulic conductivity, low infiltration rates, narrow workable moisture range, deep and wide cracks pose serious problems even at low salinity level. However, the vertisols of Baratract in Gujarat are generally very deep (150 to 200 cm), fine textured with clay content ranging from 45 to 68% with montmorillonite dominant clay minerals. The soils exhibit high shrink and swell potential and develop wide cracks of 4–6 cm extending upto 100 cm depth. The soils are calcareous in nature having calcium carbonate ranging from 2 to 12% in the form of nodules, *kankar* and powdery form. In general, they exhibit alkaline reaction. In the recent past, Sardar Sarovar irrigation project with a target to provide irrigation for about 1.8 mha of command in Gujarat has been established. The ground water quality in this region is highly saline in about 90% areas. The salinity varies from 2 to 117 dS/m with a mean of 30.7 dS/m. The use of such high salinity water either directly or in conjunction with canal water for crop production is thus limited. The salts in the sub-soil are prone to mobilize to upper surface with rising ground water table. Conditions are very favourable for secondary salinization if irrigation is practiced in traditional way. The major approach for salinity management in this region will require for prevention of rise of salinity rather than salinity reduction alone. Rainfed farming systems with *in-situ* and *ex-situ* rain water conservation, harvesting, storage, recyclingand tapping of perennial flows and augmentation of ground water for supplemental irrigation are some of the strategies for boosting the agricultural productivity of vertisols in arid and semi-arid regions. Another option is to go for biosaline agriculture in these soils. A large number of medicinal, aromatic, oil yielding and petro crops have been identified which can be cultivated with saline water irrigation[2, 3]. Some of the promising crops included *Salvadora, Matricaria, Dill (Anethum graveolens)* and grasses like *Aeloroups* and *Dicanthium*. More vigorous efforts are required to prevent irrigation induced salinity in the vertisols. Once such soils are salinized, these will require huge investment for reclamation. Vertisols need special soil and crop management practices for sustainable crop production. Cotton is the dominant crop grown in the kharif followed by sorghum and pearl millet. Pigeon pea is also grown in some area. Mostly rain fed kharif crops are grown in these area, in *rabi* season the land is either kept fallow or some fodder sorghum is grown on the residual moisture. Alternate ridges and furrows are the most commonly used layouts in the Vertisols and associated soils of India. Managing salinity of such soils not only helpful for sustainable crop production but provides food security and socio-economic development of the farming communities and hence nations.

Key References and resources for Further Reading

1. Dudal R and Bramao DL, Dark Clay Soils of Tropical and Subtropical Regions. FAO Development Paper No. 8 FAO (Food and Agriculture Organization) 1965, Rome.
2. Singh Gurbachan and Singh NT, Mesquite for the Revegetation of the Salt Lands. Bul. No. 18.1993.
3. Singh G, Singh NT, *et al.* Agroforestry in Salt Affected Soils, Bull. No. 17.1993.
4. Singh G. Salinity in Agriculture: An Overview. In Diagnosis and Management of Poor Quality Water and Salt Affected Soils (Eds Lal K, Meena RL, Gupta SK, *et al.*) Central Soil Salinity Research Institute, Karnal 2008. p1–9.

10.4 DIVERSIFIED FARMING TECHNOLOGIES FOR RURAL DEVELOPMENT IN LESSER HIMALAYAS

Agriculture is the backbone of the Indian economy and the villages are the life lines of growth of India —*Mahatma Gandhi*

Accelerated soil erosion, rapid loss of habitat, and genetic diversity are common degrading processes in Lesser Himalaya region result in low productivity of crops. Proper management of mountains, resources and socio-economic development deserves immediate strategies and management planning[1]. Crop substitution is mainly governed by production cost, productivity trends and economics. There is urgent need to select suitable crops backed by frontier technologies to substitute uneconomic and environment damaging systems on the inventory of local resources, constraints, needs and demand. Some of the important areas for diversification in the Kashmir and in some cases Ladakh region, to harness the benefits of available ecological and economic resources, include the following:

- In the unpredictable environment and weather conditions use of polyhouse technology for protected vegetable production, both in Kashmir and Ladakh region, to overcome vagaries of weather and ensure availability of vegetables round the year. Commercial floriculture under protected conditions can also accrue considerable economic advantage to the growers.
- Integrated soil fertility improvement is necessary through vermicomposting, phospho-composting, and use of biofertilisers to rejuvenate soil health. However, it still needs to be disseminate more extensively in farmers participatory mode for effective farm management and ensuring high yield.
- Specialized farming of high value and low volume crops like saffron and kala zeera and sustainable production technology for saffron cultivation need to be extended to non-traditional saffron growing areas. Similarly, kala zeera, predominantly growing in the wild need to be domesticated through standardization of agro-technology for realizing higher yields.
- Zero energy cool chambers are the on-farm storage chambers that work on the principle 'that evaporation causes cooling'. It is constructed with locally available materials that keeps the temperature 10°–15 °C less than the ambient temperature and maintains high humidity and helps for preservation of fruits and vegetables[2].
- Introduction of sunflower, as a quality oil crop and soybean as an ideal vegetable protein source particularly for livestock, since both the crops can be profitably grown in Kashmir valley.
- Integrated nutrient management strategies particularly through use of organic fertilizers in agricultural as well as horticulture crops can reduce cost of cultivation and enhance income to growers. Integrated pest management using are available technological option which can reduce the reliance of the farmers on chemical pesticides besides being environment friendly.
- Ornamental flowers of high economic value could be a profitable vocation for generation of employment and meeting national demand of flowers.
- Forage crops and natural grasses need to be intensively cultivated for dairy development in milk shed areas. Similarly, considerable feed production need also be geared up particularly maize, to help develop a potent and economically viable poultry industry. Diversification towards cultivation of fodder crops (leguminous/non-leguminous) in Ladakh region is required to make adequate fed and fodder resources available to the livestock[4].
- Diversification towards cultivation of horticultural crops with varieties of improved pedigree and performance should ameliorate the agrarian economy of the hill region. Not only the traditional fruit crops like apple, pear, cherry, almond, walnut need to be considered where improved cultivars and in some cases university developed varieties are available, but concerted efforts are needed to expand area under quality temperate fruits like kiwi, strawberry,

olive, etc. Quality of local nut types in almond and walnut need to be improved to make them competitive in the national and international market. This may require more attention in light of the state having been identified as an agri-expo zone for these crops[4].

- Popularization of other agricultural vocations, traditionally being carried out by farming community in the Lesser Himalaya, requires considerable attention both by researchers and policy planners. Small cottage industries in the rural sector in the areas of sericulture, apiculture, mushroom cultivation, sheep rearing, rabbitory, etc. have not been found lucrative enterprises for lack of adequate attention and scientific development and growth. Earnest efforts are needed to revive these sectors if this hill region is to be transformed from a subsistence level to a prosperous and sustainable agrarian economy. Second generation problems that have cropped up due to intensification of agriculture, need to be accounted for during the diversification process with innovation and creativity.
- Development of rainfed agriculture is another area of intervention. About 40% area are under rainfed and if proper varieties and resource management technologies such as mulching, water harvesting, etc. are put in place, such areas will lead to sizeable production of vegetables and fruits.
- Cultivable waste lands can optimally be used for diversified cropping systems such as agri-silviculture, silvi-pastoral and other integrated farming systems to augment the requirement of feed, fuel and fodder.

In lesser Himalaya due to small and fragmented holding of land, lack of market facilities and non availability of good planting materials hamper the success of farming community in sustainable production and result in low average productivity. Lack of agro-processing industries and transport facility coupled with less expertise and awareness doubled the problems of rural livelihoods. Remoteness and difficult terrain and severe climatic condition aggravated socio-economical development and household security in the regions. Integrated rural development approaches[3] (drinking water, education, essential goods with good infrastructure and transport facility and assured electricity needs to be created for overall sustainable rural development.

Key References and Resources for Further Reading

1. Chand S and Wani Shafiq A, Natural Resource Management in Indian Himalayas. In abstract of 9th agriculture science congress organized by NAAS and SKUAST – Kashmir, Srinagar, 22–24 June, 2009. p23.
2. Gosh P, *Curr. Sci*, 2007, 93, 1337-1338.
3. Swaminathan MS, Sustainable Agriculture: Towards an Evergreen Revolution, Konark publisher, Pvt, New Delhi 1996. India, p232.
4. Trag AR and Wani Shafiq A. Diversification for Productive Agriculture to Raise Income in Hilly Region. In souvenir of 9th agriculture science congress organized by NAAS and SKUAST — Kashmir, Srinagar 22–24 June, 2009. p61–64.

10.5 IMPORTANCE OF PRECISION FARMING

Food security is the main concern of all countries especially of the developing countries. After all the efforts made through scientific interventions in agricultural sciences seems not to be fulfilled the dreams of billions of people. Hence one has to see all the appropriate modern technologies like precision farming and resource conserving technologies.

Precision farming has been getting attention by many concerned with the production of food, feed and fiber and sustainability and efficiency of agricultural system. Swaminathan (2002) while discussing the new gains to be made towards sustainability and food security advised, that at the production level, precision farming practices will have to be developed and popularized. Alam (2007) while chairing and discussing in a brainstorming session on "Sustainable Energy for Rural India-Issues & Options" gave emphasis on the energy saving agricultural technologies through uses of data and information in the modern agriculture.

A precision farming is a management of strategy that uses information technologies to bring data from multiple sources to bear on decisions associated with crop production (Committee on Assessing Crop Yield: Site Specific Farming, Information System, and Research Opportunities, National Research Council, 1997). It has three components: data capturing, interpretation and analysis of that data, and implementation of a management response at an appropriate role and time. The key difference between conventional agricultural management system and precision farming is the application of modern information technology in micro-level decision making based on multi source data of high spatial and temporal resolution. Even though farmers know from experience that yields are higher in some part of the field than in others, conventional management focuses on input application at a given rate to the whole field. Precision agriculture on the other hand permits to apply input at a smaller scales based on new tools of leaf and soil nutrient analysis to account for the variability in the field. Variability management is the key approach in precision farming, using field maps and real time sensing for temporal and spatial variation in soils, plant growth and yield and other parameters related to cultivation (Sasao and Shibusawa, 2000).

Precision agriculture entails a paradigm shift for the agricultural research system, extension system, the farming system and infrastructure and public policy. For this, total system orientation will be required for agricultural research with close partnership of agricultural scientist, economists and information specialists. The research focus will have to shift form controlled experimental plot to on farm dynamic experimentation. The system being depended on generation of vast amount of real time data on crop variability, soil analysis has to be developed. The data so generated should be amenable to user-friendly decision support system. Understanding is the complex interactions among multiple factors affecting crop growth and farm decision-making wasted help in developing system for maximizing productivity with increased efficiency of various inputs. The farmers will become partner in research for precision farming.

Precision agricultural technologies are data intensive and geographically dispersed. Hence, a highly dispersed extension system; private or public will be required. Adequately trained professional will be required to form two bridges between the farmer and science and technology. Existing extension specialists will need continuing education and remote-site learning in precision agricultural technologies because they will have to interpret information and help the farmers take decision at farm level. Farmers owned information tools will have organized. But at the micro level a new breed of information and input service providers will possibly be the backbone of precision agriculture technology dissemination. These service providers will be entrepreneurs. It is likely that a combination of services is needed to be customized for each farmers operation. High speed data connectivity and accessibility will be the most important infrastructural support in rural areas to encourage precision farming.

As precision agriculture is information based, major policy decision on data ownership and privacy rights will have to be in place. Data collected for use at the subfield and field levels have additional value for research, testing, evaluation and marketing when assembled into regional databases. Mechanisms are needed to create and use this value, including data collection and transfer standard, institution for collecting, managing or networking data and policies to facilitate data sharing and access while protecting proprietary interests and confidentiality. As precision agriculture involves a fundamental paradigm shift, the agricultural education system has also to be reoriented to information based agriculture with direct partnership with the farmers. Precision agriculture even in the developed countries is still at the innovation stage. Unbiased, systemic, rigorous evaluations of the economic and environmental benefits and cost of precision agriculture will be required. Complexity and cost are the barriers to adoption by interested farmers even in the US, however, the economics of precision farming look to be attractive (Harris and Stanford, 1997).

A survey in Denmark, UK and the USA conducted with farmers who have adopted precision agriculture (Pedersen *et al.*, 2001) show that the farmers in general optimistic. But precision agriculture, which is actually a suit of technologies and practices, has been partially adopted by these farmers. Most of them are optimistic about the variable rate of application of fertilizers. In some cases farmers have been satisfied with solution to site specific application of pesticides saving up to 13 per cent spray (Secher *et al.*, 2000).

Indian farming situation appears to be in an advantageous position with respect to precision farming. State like Jammu and Kashmir is in introductory phase of precision farming, however even in Kashmir, farmers eagerly looking forward towards precision agriculture technologies. SKUAST-K is catalyzing farmers of Kashmir for precision agriculture technologies. Chand (2008) has been compiled 200 integrated nutrient management packages of several crops and cropping system in various agro-ecological for sustaining crop productivity and soil health in Indian conditions. Precision farming deal in small scales and as such they have more detailed information on their farms. But the organizational setup to make data acquisition, processing and decision support system available to each farmer has to be developed. In the beginning the high value cropping system should be targeted. The targeted yield equation based on basic data shall be popularized. The gain will diffuse to other systems. On the other hand for fragile agro-ecosystem like the arid ecosystem and hill-mountain ecosystem, precision agriculture can provide extensive databases and a means of controlling microorganisms to prevent deterioration. It is envisaged that a precision farming system will involve a complex network of farmers, government agencies, large information and input providers, private industries and highly skilled service-entrepreneurs. Human capital will be highly important in the system.

It is certain that precision agriculture will improve profitability through targeted and efficient application of input, which besides reducing cost, will reduce the polluting effects of inputs that do not reach the target, several studies confirmed this. For example, reduction of N_2O emission due to site-specific reduction in fertilizers application has been well documented (Munch *et al.* 2001). However, there is a possibility that precision agriculture may have less positive impact on environment than expected, for two reasons. First, reduction in fertilizer and pesticide applications may not give a proportionate reduction in their concentration. Second, increase in crop response to inputs. Similarly, better information about soil may induce farmers to increase the yield potential through more input.

Precision agriculture has the potential of making organic farming and low external input sustainable-agriculture (LEISA) feasible, provided they became profitable to both the farm service sector and the farmers. At one time it was hypothesized that precision farming wasted entail heavy expenditure but it is not so. The Government has already developed support policies in this direction mainly to increase the work and nutrient use efficiency reducing environmental reducing environmental pollution and propelling sustainable agriculture. Of course technology gap still exits and that need to be bridged. If we really want to fulfill the dreams of more than one million people of a country than, one must use the precision farming along with other conserving technologies and packages in spirit for feeding the stomach of future generation.

Key References and Resources for Further Reading

1. Alam A. Chaired and Conducted One Day Brainstorming Session on "Sustainable Energy for Rural India-Issues and Options" sponsored by NAAS—SKUAST-K, Shalimar campus on 25th October, 2007.
2. Chand, 2008. Integrated Nutrient Management for Sustaining Crop Productivity and Soil Health. International Book Distributing Co. Lukhnow, p1–142.
3. Harris D and Stanfford JV. Risk Management in Precision Farming. Precision Agriculture, 97, Vol. II: Technology, IT and Management, Papers presented at the First European Conference on Precision Agriculture, Warwick University, UK 7-10 September 1997, p851–859.

4. Sasas A and Shibusawa S 2000. Prospects and Strategies for Precision Farming in Japan. Japan Agricultural Research Quarterly, 34:233–238. Committee on Assessing Crop Yield-Site Specific farming, Information Systems and Research Opportunities, National Research Council (US) 1997. Precision Agriculture in the 21stCenturay: Geospatial and Information Technologies in Crop Management. Pub. National Academy Press (US) p149.
5. Secher BJM, Bjerre KD and Seiero M 2000. Site Specific Control of Pest and Diseases — A Challenges and An Opportunity. The BCPC Conference: Pest and Diseases, Vol. II: Proc. International Conference, Brighton, UK, p13–16.
6. Swaminathan MS 2002. Food Security and Sustainable Development. In Acharya SS, Singh Surjit and Sagar Vidya (ed), Sustainable Agriculture, Poverty and Food security: Agenda for Asian economics Vol. I p11–32.
7. Munch JC, Berkenkamp A, Sehy U 2001. The Effect of Site Specific Fertilization on N_2O Emissions and N-Leaching Measurement and Simulation. In Hort WS *et al.* (ed), Plant Nutrition: Food Security and Sustainability of Agro-ecosystems Through Basic and Applied Research—Fourteenth International Plant Nutrition Colloquiums, Handover, Germany, p102–903.

10.6 ORGANIC FARMING AND CERTIFICATION

Organic farming is defined worldwide as farming system without the addition of synthetic chemicals that is the ones that have been manufactured or have been processed chemically, e.g. fertilisers, pesticides, growth regulators, etc. The definition of 'organic agriculture' includes the word *addition* because organic farming is not necessarily chemical free farming. This is because we live in a world where there are artificial chemicals in the soil, air and water. It combines the best of old knowledge and traditions with the best of modern sciences. It has much in common with other good management. Organic farms, like other well run farms; need a high level of management, particularly of soils and pest. Organic farming is a system approach blending indigenous knowledge with modern scientifically proven technology utilizing the naturally recycled farm by-products and nutrient inputs of biological interactions for crop production and protection. In India organic movement started in Madhya Pradesh in 2001 and then spread all over India. Throughout the history, the focus of agricultural research and majority of published scientific findings has been shifted to biotechnologies like genetic engineering. The rise of organic farming was driven by small, independent producers and by consumers. In recent years, explosive organic market growth has encouraged the participation of agribusiness interest. Organic farming is a holistic production management system based on basic principle of minimizing the use of external inputs and avoiding the use of synthetic fertilizers and pesticides to ensure sustainability of agriculture. There are several regions in the country where organic farming is being practiced by farmers and it is becoming increasingly popular throughout the country. Although, presently organic farming is practiced in about 43000 hectares of farm land constituting only 0.03% of total cultivable land, but there is a large potential for organic farming in India due to extensive cropped area and wide range of agro-climatic conditions[3, 4]. Availability of huge amounts of agricultural waste and alternative sources of nutrition like vermicompost, biofertilisers, etc. can offer a better scope of organic agriculture. The government of India has also initiated several programmes to promote organic farming including identification of potential areas and target crops especially the crops having export potential. Priority zones have been identified based on cropping pattern, rainfed irrigation and fertilizer use, etc. But lack of proper market infrastructure and distribution network is acting as a major bottleneck in growth of organic agriculture[2]. With the efforts of government to streamline regulatory mechanisms for export of organic produce and awareness among local population for domestic consumption will pave way for faster development of organic farming, a realistic approach to sustainable agriculture and healthier food.

Certification means having the farm and the farmer's methods inspected by an accredited agency to ensure that they comply with the guidelines on the organic farming. Each certifying agency has a code of standard to prevent the marketing of substandard produce. Different certifiers use different inspection methods and criteria but the results are similar. Some certifiers use three levels of organic certification viz. Level A or the top level, is fully organic level; Level B, or the in-conversion level, is the transitional level, for produce from farms, which are being converted to organic farms. Farms in this level must meet the level A standard but are not considered organic until they have been farmed in this way for sometime–usually at least 2 years. And Level C or the pre-conversion level, is the period prior to the in-conversion level. The farm and its operations are under an organic inspection system, usually for a period of 12 months.

Certification of organic farms by the accredited agency is mandatory for export. After deciding to have your farm certified, commit yourself to meeting the standards of the certifying group and resisting the temptation to reach for a chemical when things get tough. In India, Ministry of Commerce has designed national standards for organic products which could be sold as "India Organic". Agriculture and processed food products export development authority (APEDA), Coffee Board, Tea Board, Spices Board are appointed accreditation agencies in India. APEDA has also recognized a couple of laboratories for export testing like Sri Ram Institute for Industrial Research, New Delhi and Food Research and Analysis Center, New Delhi. M/s international resources for fairer trade (IRFT), Mumbai is certified under national standards of organic programme and accredited by APEDA. Besides this, there are currently ten certifying agencies in India providing certification under National Standards of Organic Production (NSOP)[1].

Farmers should start with small area to discover limitations and identity possible problems to address the constraints in term of organic farming. One has to identify markets for the produce, and decide how and when he is going to market the produce, e.g. whether to export, supply to organic wholesaler, processor or sell at farmer market or at the farm gate. The requirements of such markets (e.g. quality and packaging) for product processing so as to add value to the produce, promotion of the product and likely price expected. Organically grown crops may not initially reach the crop yield levels of integrated management fields or mineral source fertilized fields. The farmers must have the financial reserves to cover the drop in income, which usually occurs during conversion to organic farming.

Key References and Resources for Further Reading

1. Bhattacharyya P and Verma VK In: Souvenir: National Seminar on National Policy on Promoting Organic Farming. *National Centre of Organic Farming, Ghaziabad* 2005. p40–43.
2. Dwivedi V, In: Souvenir: National Seminar on National Policy on Promoting Organic Farming. *National Centre of Organic Farming, Ghaziabad* 2005, p58–61.
3. Subba Rao A, Chand S and Srivastava S *Fertil. News* 2002. 42:75–95.
4. Chand S and Somani LL, *Int. J. Tropical Agriculture* 2003. 21:133–140.
5. Worthington V, *J Alternative and Complimentary Medicine* 2001. 2:161–173.

10.7 GUIDELINES FOR MAXIMIZING FERTILIZER USE EFFICIENCY

Fertilizers are main component for high yield sustainable agriculture. Choice of a fertilizer depends on unit cost of nutrients present in it and its agronomic efficiency under a given situation. Fertilizer is a valuable input and measures should be taken to reduce its losses and to increase its uptake and utilization by the crop. Selecting a situation-specific fertilizer and choosing the time and method of application according to crop demand would minimize losses and increase its efficiency.

N-Fertilizer

Most crop plants recover only 25%–35% of the nitrogen applied as fertilizers. Losses occur by ammonia volatilisation, denitrification, and immobilization to organic forms, leaching and run off. Utmost care should be bestowed in selecting the type of fertiliser as well as the timing and method of application. The right choice of fertilizer and right dose alongwith right time of application leads a bumper maize crop (Fig. 10.2).

Fig. 10.2 Bumper maize crop in field due to best fertilizer management practices

Choice of Nitrogenous Fertiliers

- In submerged rice soil, ammoniacal and ammonia-producing fertilizers like urea are most suitable since ammonia is the most stable form of nitrogen under such conditions.
- For acidic upland soils, ammoniacal fertilizers are most suitable during rainy season since ammonium is adsorbed on soil particles and hence leaching losses are reduced. Adsorbed ammonium is gradually released for nitrification and thus becomes available to crops for a longer period.
- In highly acidic upland soils, urea is preferred to ammonium sulphate as the former is less acid forming.
- In alkaline upland soils of low rainfall regions, nitrate fertilizers are preferred to ammoniacal fertilizers or urea since ammonia may be lost by volatilization under alkaline conditions.

Management of Nitrogenous Fertilizers

- Almost all the nitrogenous fertilizers are highly amenable to losses and since most of the crops require nitrogen during the entire growth period, split application is necessary to ensure maximum utilization by crops.
- More number of splits may be given for long duration crops as well as perennial crops.
- Nitrogen losses from fertilizers are more in coarse textured soils with low cation exchange capacity (CEC) than in fine textured soils. Hence more number of splits are necessary to reduce loss of fertilizer nitrogen from sandy and other light soils.
- For medium duration rice varieties, nitrogenous fertilizers should be given in three splits, as basal, at maximum tillering and at panicle initiation stage.
- In coarse textured sandy or loamy soils, the entire dose of nitrogenous fertilizers may be applied in 3–4 splits at different stages of growth of rice crop.
- In areas where split application of nitrogen is not feasible due to water stagnation after planting/sowing, full dose of nitrogen as basal may be given in the form of neem coated or coal tar coated urea.
- In double-cropped wetlands, 50% of N requirement of the first crop may be applied in the organic form.
- As far as possible, liming should be done one or two weeks prior to the application of ammoniacal or ammonia forming fertilizer like urea since ammonia is likely to be lost by volatilization if applied along with lime.
- Almost 70% of N in urea applied by broadcast to flooded soil is lost by volatilization, immobilization and by denitrification.

Measures to Reduce the Loss of Nitrogen from Applied Urea

- Urea supergranules or urea briquettes may be used in places where soil is clayey and has cation exchange capacity more than 10 cmol (+) per kg of soil.

- Sulphur or lac coated urea is suitable where soil is liable to intermittent flooding and in situations where water management is difficult. This is more suitable for direct sown crop.
- Urea may be mixed with moist soil and kept for 24–48 hours before application to the field. Alternatively, urea may be mixed with moist soil, made into balls of about three inch diameter and dried under shade. The balls may be placed deep into subsoil.
- Mixing urea with five times its weight of neem cake prolongs the period of nitrogen availability to the crop.
- For submerged soils, coating urea with coal tar and kerosene (100 kg urea is mixed with 2 kg coal tar dissolved in one litre kerosene) before mixing with neem cake is preferred to simple mixing with neem cake.
- Coating urea with neem extract (containing about 5% neem triterpenes) at 1% rate and shade-drying for 1 to 1.5 hours before applying in direct-seeded puddled lowland rice increases nitrogen use efficiency.
- As far as possible, urea may be applied by deep placement or plough sole placement. Deep placement of prilled urea or super granules during the last ploughing followed by flooding and planting is beneficial in light soils. Urea briquettes or super granules may be placed between four hills of transplanted rice, whereas sulphur coated or lac coated urea may be broadcast on the surface.
- Foliar spray of 5% urea solution can be practised in situations where quick response to applied nitrogen is required. If power sprayers are used, the concentration may be increased to 15%. Fresh urea should be used to avoid toxicity due to biuret.
- Awareness and training to farmers helps in dissemination of knowledge about fertilizer use and management in modern farming system.

P-Fertilizers

Fertilizer phosphorus is an expensive input and its management poses serious problems due to several complexities in its behaviour in different types of soil. This often results in its poor recovery from applied fertilizers.

Choice of P-Fertilizers

- In slightly acid, neutral or mildly alkaline soils, water-soluble phosphatic fertilizers are more suitable.
- In wetland rice soils, water-soluble phosphatic fertilizers are preferable as pH of most of the submerged soils is near neutral.
- In strongly acidic soils whose pH does not rise above 5.5 to 6.0 even on submergence, phosphatic fertilizers containing citrate soluble form of P like basic slag, dicalcium phosphate, steamed bone meal etc. are suitable.
- For highly acidic upland soils or submerged soils whose pH will not rise above 5.5 even on submergence, powdered rock phosphate is suitable. Soil acidity

converts tricalcium phosphate in rock phosphate to plant available monocalcium form.

- For short duration crops where quick response is required, water-soluble phosphatic fertilizers are most suitable.
- For perennial crops like rubber, oil palm, coffee, tea, cardamom, etc. phosphorus in the form of rock phosphate can be applied.
- In black soil phosphatic fertilizers containing water-soluble phosphate like single superphosphate are most suitable.

Management of Phosphate Fertilizers

- Acid soils have to be amended with lime, dolomite or magnesium silicate and alkali soils with iron pyrite or sulphur before application of phosphatic fertilizers. This will help to reduce fixation and increase availability of P.
- Surface application or broadcasting is preferred for shallow rooted crops whereas placement in the root zone is advantageous in deep rooted crops.
- Rock phosphates can be used advantageously in rice grown in acid soils during the *virippu* season. Powdered rock phosphate may be applied and mixed thoroughly with soil by ploughing. After two or three weeks, the field may be flooded, worked up and planted with rice. Under this situation, phosphorus in rock phosphate gets converted to iron phosphate, which on subsequent waterlogging becomes available to the rice crop.
- Rock phosphate can be used successfully as a phosphatic source for leguminous crop since its root system can extract phosphorous from rock phosphate.
- In single crop wetlands where rice is grown in the *virippu* season, application of phosphatic fertilizers can be dispensed with for the rice crop, if the second crop (usually legume or green manure) is given phosphatic fertilizers.
- In case of rice-legume cropping sequence in acid soils, application of rock phosphate to the pulse crop helps to skip phosphatic fertilizers in the succeeding rice crop.
- Since phosphorus requirement of seasonal crops is confined to the early stages, phosphatic fertilizers are to be applied at the time of seeding or planting.
- Top dressing of phosphatic fertilizer leads to wastage of the fertilizer nutrient. Further, excessive phosphates may lead to deficiency of micronutrients such as zinc, boron, etc.
- Under adverse soil conditions and where quick result is required, spraying water-soluble phosphatic fertilizers like triple superphosphate or hot water extract of superphosphate can be resorted to.

Fig. 10.3 Awareness among farmers must be created through training and workshop (*Author interacting farmers during training*)

K-Fertilizers

For most crops, potassium can be supplied as muriate of potash. But in crops like tobacco and potato, muriate of potash may cause chloride injury, reducing quality of the produce. In such cases, K may be applied as potassium sulphate.

Management of K-Fertilizers

- In coarse textured soils and in heavy rainfall regions, potassium fertilizers should be applied in as many splits as possible, to reduce loss of potassium.
- In fine textured soils, the entire dose of potassium fertilizers may be applied as basal.
- In acid soils, potassium fertilizers should be applied only after lime application to prevent loss of potassium by leaching.

Conclusion

Besides major nutrient, it is essential to use secondary and micronutrient fertilizer mixture for maximum benefits from fertilizer. Wise use of fertilizer not only increases the crop productivity and water but reduces cost of cultivation. Best fertilizer management practices (BMPs) are important for sustainable crop productivity and soil health in modern farming system where nutrient turnover is higher.

10.8 SOIL POLLUTION THROUGH AGROCHEMICALS AND THEIR REMEDIAL MEASURES

An ever-growing world population increasing pressure on agriculture to meet the human needs of food, fiber, oils etc. Modern agriculture has brought a drastic change leading to rather a dynamic inorganic chemical dependent agriculture which was a few decade back a static soil dependent. With advancement in irrigation resources coupled with 110-fold increase in fertilizer consumption since 1950 and correspondingly increase in food grain production 51 to 204 million tones in 2004[1]. With this modernization in agriculture, our country achieved its goal of self sufficiency in food production but has resulted in its value the exploitation of natural resources.

As high as 50 per cent of our food is lost due to pest and parasites the way of hope left in modern agriculture is to use pesticides (insecticides, fungicides, nematicides, antibiotics, rodenticides). Further, the use of pesticides have not only restricted to agriculture, rather it has entered in the day to day life of a common man. Soil, air, water pollution is at present day concern. The toxic element added through various chemicals to the soil may either be absorbed by the growing plants or pass on to the underground water. When absorbed quantities of these pollutants exceed certain limits, may become hazardous not only to human beings and animal but to the present day economy as a whole.

Soil pollutants: The soil is primary, intended or otherwise of many of the waste products and chemicals used in the modern society. Six general pollutants commonly reach the soil, viz. pesticides, inorganic pollutants, organic pollutants, organic wastes, salts, radio-nuclides and acid rain are described below:

Pesticides: Chemicals being used to control harmful organisms are known as pesticides. The chemical revolution in agriculture began with the discovery of the insecticidal properties of DDT in 1939. Pesticides are not readily biodegradable, It is believed that as little as one per cent of the pesticides applied come in contact with the targeted organisms. The residue left in the soil may adversely affect the soil flora and fauna including fish and other animals. The problem of pollution doubled when the pollutant enter the food-chain such as birds, fish, being secondary and tertiary consumers tend to concentrate these chemicals in their body tissue beyond the toxic limits.

Inorganic Pollutants: There are many sources of inorganic chemical contaminants that can accumulate in the soil. The burning of fossil fuels, smelting and other processing techniques release tons of these elements into the atmosphere which can be carried for miles and later deposited on the vegetation and soil. Lead, nickel, and boron are gasoline additives that are released into the atmosphere and carried to the soil through rain and snow. Superphosphate and limestone usually contain small quantities of cadmium, copper, manganese, nickel and zinc. Arsenic for many years used as an insecticide on cotton, tobacco and fruit crops. Whatever their sources, toxic elements upon and do reach in the soil, where they became part of life cycle of soil-plant-animal-human.

Fertilizers on one hand are responsible for higher food production but another hand known to be toxic pollutants. Added fertilizer into the soil subjected to various losses i.e. dentrification, leaching, volatilization-action, etc. especially in soluble fertilizer. Elevated concentrations of nitrates in drinking water, particularly when it is contaminated with bacteria, have been associated with causing blue baby disease (*Methaenoglobinaemia*) in infants and babies. Nitrate has been linked as a higher risk factor causing various kinds of cancer in older persons due to formation of *Nitrosamines* in the digestive tract. Continuous use of phosphatic fertilizer is responsible for cadmium build up into soil. An excess of potassium in grassland accompanied with higher applications of nitrogen may lead to grase tetany (*hypomagnesmia*) in cattle.

Organic Waste: The domestic and industrial sewage sludge is the major sources of toxic elements. Although organic wastes exert beneficial physical effects on soil and are also the source of plant nutrients but they also carry significant quantities of inorganic as well as organic chemicals leading to soil and environmental degradation. Application of sewage and sludge containing toxic quantities of heavy metals is harmful to earthworms growing in the soil. Heavy metals in soils under normal conditions are present in traces but the fall-out from metal smelters markedly, increases their levels in the soil.

Salts: Contamination of soil with salts is also a form of soil pollution. Mostly salt accumulates in the soils of arid and semi-arid regions where evapotranspriation exceeds precipitation and soil parent material containing high concentration of salt. Another reason may be of irrigation with salt-laden irrigation water and hard pan in the subsurface. The control of salinity depends almost entirely on the quality and management of water. Sulphur and gypsum application can be used to eliminate toxic sodium bicarbonate. Drainage is another control measure of salts.

Radio-nuclides: Nuclear fission in connection with atomic weapons testing provides another source of soil contamination. To the naturally occurring radio-nuclides in soil a number of fission products have been added. Only two of these are sufficiently long lived to be of significance in soils, strontium-90 (half-life is 28 yr) and cesium-137 (half-life is 30 yr). Besides from these nuclei low level radioactives (plutonium, uranium, curium, etc.) are also responsible for soil pollution. The soil is the primary source of radon, a colorless and odorless radioactive gas that can cause lung cancer. Radon is formed from the radioactive decay of radium. Radon enters homes and other buildings from surrounding soils.

Acid Rains: Acid precipitation popularly called acid rain is apparently due to the oxidation of nitrogen and sulphur containing gases that dissolve in the water vapor of the atmosphere to form nitric and sulphuric acid-reactions such as:

$$2NO + O_2 \rightarrow 2NO_2 + H_2O \rightarrow HNO_3 + HNO_2$$
$$2SO_2 + O_2 \rightarrow 2SO_2 + 2H_2SO_4$$

It is serious for soils that are already quite acidic since increased acidity will make them even less fertile.

Preventive Measures

Reducing soil application: First, action is required to reduce unintentional aerial contamination from industrial operations and from automobile, truck, and bus exhaust. Decision makers must recognize the soil as an important resource that can be seriously damaged by contamination from accidental addition of inorganic toxins, such contamination must be curtailed. Also, there must be judicious reductions in application of the toxins through pesticides, fertilizers, irrigation water, and solid waste.

Reducing recycling: Soil and crop management can help in reducing the continued cycling of toxic inorganic chemicals. This is done primarily by keeping the chemicals in the soil and reducing their uptake by plants. The soil becomes "sink" for the toxins[2]. The cycle is broken by immobilizing the toxins in the soil. For example, most of these elements are rendered less mobile and less available if the soil pH is kept near neutral or above. Liming of highly acid soils should also be beneficial, since the oxidized forms of the toxic elements are generally less soluble and less available for plant uptake than the reduced forms. Heavy phosphate application reduces the availability of toxic cations but may have the opposite effect on arsenic, which is found in the anionic form, leaching may be effective in removing excess boron, although moving the toxin from the soil to nearby waterways may not be of any real benefit. Advantage can be taken of differences in the abilities of crop species or varieties to extract the toxins. Even though none of the crops tested take up a high proportion of the elements, some absorb more than others, "accumulators" should be avoided if the harvests are to be fed to human or domestic animals. Moreover, forage crop should be harvested at the stage of maturity when the concentration of the toxin is lowest. It is obvious that soil and crop management offers some potential for alleviating contamination by inorganic elements.

Landfills: Soils have long been used as disposed "sinks" for municipal refuse similarly "landfills" are widely employed to dispose of a variety of wastes from our towns and cities. These wastes include paper products, garbage, and non-biodegradable materials such as glass and metals. The sites are often located in swampy lowland areas that eventually become built up by the dumping to create upland areas for such uses as city parks and other facilities. Unfortunately, sanitary landfills are sometimes not so sanitary. Through leaching and runoff these sites can contaminate both surface and ground water. Contaminants include heavy metals as well as soluble and biodegradable organic materials. The environmental hazards associated with landfills require marked restriction on their continued use.

Organic matter: Soil organic matter can adsorb trace elements pollutants (e.g. Pb, Cd, Cu) which will reduce their chances of contamination of surface and ground water[3]. Another advantage is in the adsorption of pesticides and other organic chemicals. Organic matter reduces the possibility of pesticides carryover effects, prevents contamination of environment and enhances both biological and non-biological degradation of certain pesticides and organic chemicals. In addition organic matter is known for its capacity to adsorb inorganic (e.g. $NO + NO_2$) and organic gases (e.g. CO). Addition of early decomposed organic matter is an important practice to reduce pesticide level in soils. The use of high nitrogen containing cover crops and their addition in large quantities and animal manures are also helpful in pesticide level reduction. Organic recycling promotes microbial action to facilitate degradation of even the most resistant pesticides. Biochemical degradation of pesticides by soil organisms is the single most important method by which pesticides removes from the soil. Altering soil pH, increasing sorption capacity, precipitation of trace elements as some insoluble solid phase, attenuation, metal-P interaction and enhancing volatilization losses can reduce toxic levels of metals in soil.

Hyper accumulator: These plant species are able to accumulate metals in their tissues in very high concentration. *Thlaspi alpester* is a zinc hyper accumulator that has been found growing when concentrations were extremely high. Biomass production by *Thlaspi* is very low and it is hoped that this can be improved with biotechnology to the point that zinc removal from soils-water such a hyper accumulator might be feasible.

Phyto-remediation: There are several bioremediation techniques that may be feasible for integration with phytoremediation strategies for enchanting contaminant reduction. Incorporation of appropriate amendments (biostimulation) may be used prior to introduction of plants to initiate degradation and reduce toxicity of contaminants. Land farming or dilution of the contaminant in soil followed by appropriate plant establishment may be useful in reducing the initial toxicity. Surfactants may have utility to enhance solubility of xenobiotics may facilitate plant uptake. However, selection of plants that exude appropriate biosurfactants through their roots may be a superior strategy for enhancing rhizodegradation. The complex nature of soil contamination requires numerous management considerations related to phytoremediation.

A Healthy Human Life Depends on Healthy Soils and Quality Produces,
For that I Appeal to Save Nature, Soil, Water, Energy and Resources
For Present as well as Future Generations of Mankind
Keep Green and Clean Earth Planet

—Chand Subhash

Important Terms

AC soil: A soil having a profile containing only A and C horizons with no clearly developed B horizon.

Acid rain: Atmospheric precipitation with pH values less than about 5.6, the acidity being due to inorganic acids such as nitric and sulphuric that are formed when oxides of nitrogen and sulphur are emitted into the atmosphere.

Acid sulphate soils: Soils having sufficient sulphides (FeS_2 and others) and pH < 3 and when drained and aerated enough for cultivation also termed as cat clays.

Acidity, active: The activity of hydrogen ion in the aqueous phase of a soil. It is measured and expressed as a pH value.

Acidity, residual: Soil acidity that can be neutralized by lime or other alkaline materials but cannot be replaced by an unbuffered salt solution.

Acidity, salt replaceable: Exchangeable hydrogen and aluminum that can be replaced from an acid soil by an unbuffered salt solution such as KCl or NaCl.

Acid soil: A soil that is acid in reaction throughout the root zone. Practically, this means a soil having pH less than 6.6; precisely, a soil with a pH value less than 7.0. Such a soil has more hydrogen (H) than hydroxyl (OH) ions in the soil solution. Acid soils are grouped into five categories.

Extremely acid	:	pH below 4.5
Very strongly acid	:	4.5–5.0
Strongly acid	:	5.1–5.5
Medium acid	:	5.6–6.0
Slightly acid	:	6.1–6.5

Acid soils are reclaimed by addition of liming materials.

Acidity, total: The total acidity in a soil. It is approximated by the sum of the salt replaceable acidity plus the residual acidity.

Acid-forming: A term applied to commercial fertilizers which leave an acid residue in the soil. The amount of calcium carbonate required to neutralize the acid residue is referred to its equivalent acidity. Common fertilizers along with their equivalent acidity. Common fertilizers along with their equivalent acidity.

Alkali Soil: A soil that contains sufficient sodium salt to interfere with the growth of most crop plants.

Alkaline-forming: A term applied to commercial fertilizers that leave an alkaline or basic residue in the soil. The basic residual effect is expressed in terms of 'equivalent basicity'. More the basic residue more will be the equivalent basicity.

Alkaline soil: A soil that is alkaline in reaction throughout the root zone or for a major part of the root zone. Precisely, any soil having a pH value greater than 7.0. Practically, a soil having a pH above 7.3 is called alkaline. An alkaline soil is reclaimed by addition of gypsum or sulphur. Soils having pH from 7 to 7.5 are sometimes referred to as alkali soils. They correspond to black alkali soils and occur in irregular patches. They can be reclaimed by adding calcium or magnesium.

Allophane: An aluminosilicate mineral that has an amorphous or poorly crystalline structure and is commonly found in soils developed from volcanic ash.

Alluvial soil: A soil developing from recently deposited alluvium and exhibiting essentially no horizon development or modification of the recently deposited materials.

Alluvium: A general term for the detrital material deposited or in transit by streams, including gravel, sand, silt, clay and all variations and unconsolidated mixtures of these.

Amendment, soil: Any substance other than fertilizers, such as lime, sulphur, gypsum, and sawdust, used to alter the chemical or physical properties of a soil, generally to make it more productive.

Biogas: The gas produced from organic waste by microbiological reaction under anaerobic conditions. The organic wastes used in the production of biogas are generally cattle yard waste, human waste, vegetative crop residues, unwanted aquatic plants and weeds. Biogas is sometimes termed after the material or method of gas production, e.g. marsh gas, sewage gas, sludge gas, digester gas and gobar (cowdung) gas, etc. As a fuel, biogas is a potential source of energy. It is usually with the following composition.

Methane	=	50%–60%
Carbon dioxide	=	30%–40%
Hydrogen	=	5%–10%
Hydrogen sulphide	=	traces
Water vapour	=	traces

Biological test: A technique for assessing nutrient status of soil using biological material. The following types of tests are commonly used:

1. Field test: Involves use of field crops in experimental or farmers' fields.
2. Laboratory or greenhouse tests:
   a. Mitscherlich pot culture
   b. Lettuce pot culture
   c. Neubauer seedling method
   d. Sunflower pot culture
3. Microbiological tests:
   a. *Azotobacter* culture
   b. Sackett and Stewart technique
   c. *Aspergillus niger* test
   d. Mehlich's Cunninghamella plaque method

Biomass: The amount of living matter in a given area.

Biuret ($C_2O_2N_3H_5$): A chemical compound formed by the combination of two molecules of urea with the release of a molecule of ammonia, when the temperature during the urea manufacturing process exceeds a certain level. Fertiliser grade urea contains variable amounts of biuret. Biuret is toxic to plants, particularly when it is applied through sprays. As per the Indian Fertilizer legislation (FCO, 1985) the biuret content in urea should not exceed 1.5%.

Buffering capacity: The ability of a soil to resist changes in pH. Commonly determined by presence of clay, humus, and other colloidal materials.

Bulky organic manures: These manures are bulky in nature and supply plant nutrients in small quantities and organic matter in large quantities.

Calcareous soil: Soil containing sufficient calcium carbonate (often with magnesium carbonate) to effervesce visibly when treated with cold 0.1 N hydrochloric acid.

Carbon: An essential plant element. Plants take their carbon requirement from atmospheric carbon dioxide. Since it is never deficient in the atmosphere to become a limiting factor in growth, it is not supplied in any form as fertilizer. Carbon present in organic compounds in soil cannot be used as a nutrient source of carbon.

Carbon cycle: The sequence of transformations whereby carbon dioxide is fixed in living organisms by photosynthesis or by chemosynthesis, liberated by respiration and by the death and decomposition of the fixing organism, used by heterotrophic species, and ultimately returned to its original state.

Carbon-nitrogen ratio: The ratio of the weight of total organic carbon to the weight of total nitrogen in a soil or in an organic material. C—N ratio of wheat straw is nearly 80:1 while that of soil is 10:1. When undecomposed straw with a high C—N ratio is applied to the soil, its C—N ratio is reduced through bacterial decomposition.

To speed up bacterial decomposition, nitrogenous fertilizer is added to the soil at the time of turning under straw.

Cation: An ion carrying positive charge of electricity. The common soil cations are calcium, magnesium, sodium, potassium and hydrogen.

Cation exchange: The exchange of cations held by the soil absorbing complex is rich in sodium (as is the case in alkali and alkaline soil), application of gypsum (calcium sulphate) causes calcium cations to exchange with sodium cations.

Cation exchange capacity: The sum total of exchangeable cations that a soil can adsorb. Sometimes called "total-exchange capacity" "base-exchange capacity", or "cation-adsorption capacity." Expressed in centimoles per kilogram (cmol/kg) of soil (or of other adsorbing material such as clay).

Chelates: An organic compound capable of holding the plant nutrient in a form which prevents it from getting tied with other elements in the soil, thus keeping it more or less in available form for the plant. The term refers to the claws of a crab illustrative of the way in which the atom is held. For examples EDTA, DTPA, HEEDTA and CDTA.

Diatomaceous Earth: A geologic deposit of fine, grayish, siliceous material composed chiefly or wholly of the remains of diatoms. It may occur as a powder or as a porous, rigid material.

Diatoms: Algae having siliceous cell walls that persist as a skeleton after death, any of the microscopic unicellular or colonial algae constituting the class Bacillariaceae. They occur abundantly in fresh and salt waters and their remains are widely distributed in soils.

Dry farming: The system in which field crops are raised with an annual rainfall of less than 25 inches, without irrigation facilities. From plant nutrition point of view, lesser doses of fertilizers are recommended to dry farming areas for all field crops, compared to irrigated farming.

Dryland farming: The practice of crop production in low rainfall areas without irrigation.

Efficacy of fertilizer: The whole of the amount of nutrient from the applied fertilizer is not recovered by the crop as a part of it is lost or remains in the soil. The extent of recovery of applied nutrient by a crop or crop rotation indicates the efficacy of a fertilizer. Some factors for controlling efficiency of fertilizer application are listed below:

- The nature of crop and its variety
- Method and time of application of fertilizer
- Crop management
- Cropping system

- Chemical composition of soil and its pH
- Organic matter content of the soil
- Physical condition including drainage, aeration, etc.
- Weather conditions
- Soil moisture
- Balance of nutrients

Earth flow: The process of saturated soil moving down a slope under the force of gravity. The term is also used to describe the results of the process.

Effluent: Liquid waste from either a septic tank or a sewage treatment plant.

Eutrofication: Pollution with unwanted nutrients.

Geographical information system (GIS): It is a computerised data base management system for capturing, storing, validating, analysing, displaying and managing spatially referenced data sources in addition to the primary data such as agro-climatic and soil characteristics. It can be used to generate map on erosion hazard, land suitability for a specified alternative land use type.

Greenhouse effect: The warming of the earth's surface and atmosphere owing to absorption of out-going radiation by CO_2, CH_4 and H_2O (like absorption by glass).

Heavy soil: A soil with a high content of the fine separates, particularly clay, or one with a high drawbar pull, hence difficult to cultivate.

Hematite, Fe_2O_3: A red iron oxide mineral that contributes red color to many soils.

High-analysis fertilizer material: The fertilizers containing a higher percentage of the plant nutrients. Such as urea containing 46 per cent nitrogen, triple superphosphate containing 45 to 47 per cent P_2O_5 and muriate of potash containing 60 per cent K_2O. High-analysis fertilizers have an advantage in the cost of bagging, handling and transportation per unit of plant.

Infrared: Refers to the electromagnetic radiation of wavelength longer than light but shorter than radiowaves. Sometimes called heat radiation.

Integrated plant nutrition system (IPNS): It is a concept proposed by FAO where the basic goal is the maintenance or adjustment and possibly improvement of soil fertility and of plant nutrient supply to an optimum level for sustaining the desired crop productivity through optimization of the benefits from all possible sources of plant nutrients in an integrated manner.

Light soil: A term used for sandy or coarse-textured soil. Leaching of nutrients is more in light soils, as such it is advisable to apply nitrogenous fertilizers in small doses. Wherever practicable, green manuring should be done to improve the physical condition of a light soil.

Lime (agricultural): In strict chemical terms, calcium oxide. In practical terms, a material containing the carbonates, oxides and/or hydroxides of calcium and/or magnesium used to neutralize soil and acidity.

Liquid fertilizers: Commercial fertilizers in liquid form. Such fertilizers are chiefly anhydrous ammonia, aqueous solution of nitrogen, and some mixed fertilizers. Liquid fertilizers are applied to the soil, through irrigation water, starter solutions, or with the help of special equipment. Use of liquid fertilizers is common in the United States of America.

Lithosequence: Two or more soils in which the soil forming factor that varies the most is the parent material. Other soil-forming factors are constant or vary much less than parent material.

Longwave radiation: In meteorology, radiation in the infrared and radio wavelengths, emitted at the earth's surface and partly absorbed in the atmosphere. Distinct from the sun's shortwave radiation.

Low-analysis fertilizer materials: The fertilizers containing a low percentage of plant nutrients. Such fertilizers are Chilean nitrate containing 16 per cent nitrogen and single superphosphate containing 16 per cent nitrogen and single superphosphate containing 16 to 18 per cent P_2O_5.

Manure: The excreta of animals — dung and urine, with straw or other materials used as the absorbent. The decomposed manure is called farm yard manure or farm manure or barn yard manure. The average composition of well-rotted farm yard manure is 0.5 per cent nitrogen, 0.3 per cent P_2O_5 and 0.5 per cent K_2O.

Methane, CH_4: An odorless, colorless gas commonly produced under anaerobic conditions. When released to the upper atmosphere, methane contributes to global warming.

Micron (μ): A unit of length denoting a millionth part of a meter $1\ \mu = 10^{-6}$ meter. In instrumental analysis it is used to measure wavelength of radiation.

Microfauna: That part of the animal population which consists of individuals too small to be clearly distinguished without the use of a microscope. Includes protozoans and nematodes.

Microflora: That part of the plant population which consists of individuals too small to be clearly distinguished without the use of a microscope. Includes actinomycetes, algae, bacteria, and fungi.

Neem coated urea: Neem has a nitrification inhibition property, hence is used in India for coating of urea granules to produce slow release fertilizer. The technique is developed at the Indian Agriculture Research Institute for coating urea with neem cake is as follows: About 100 kg urea is mixed in a drum with a solution of

1 kg coaltar in 2 litres of kerosene. To this, 20 kg of powdered neem cake is added and thoroughly mixed.

Neutral soil: A soil in which the surface layer, at least to normal plow depth, is neither acid nor alkaline in reaction. In practice this means the soil is within the pH range of 6.6–7.3.

Night soil: Night soil is human excreta, solid and liquid, In India, it is directly applied to the soil to a limited extent but converted mainly as town compost. In cities which have sewage facilities, sewage water and sludge are used directly to raise crops. On an average right soil contains 5.5 per cent nitrogen, 4.0 per cent phosphorus (P_2O_5) and 2.0 per cent potash (K_2O) on oven dry basis.

Niter (KNO_3): Potassium bearing mineral also known as potassium nitrate or saltpeter. It contains about 46% K_2O.

Nitrate of soda: Commercial nitrogenous fertilizer carrying 16 per cent nitrogen in nitrate form. Nitrate of soda is imported from Chile hence the name Chilean nitrate. In other countries, it is manufactured synthetically.

Nitrification: The formation of nitrates and nitrites from ammonia (or ammonium compounds) as in soils, by microorganisms.

$$NH_3 \xrightarrow{\text{Nitrosomonas}} NO_2 + 3H^+$$

$$NH_3 \xrightarrow{\text{Nitrobactor}} NO_3$$

Nitrogen cyle: The sequence of chemical and biological changes undergone by nitrogen as it moves from the atmosphere into water, soil, and living organisms, and upon death of these organisms (plants and animals) is recycled through a part or all of the entire process.

Organic farming: According to the US Department of Agriculture "A production system which avoids or largely excludes the use of synthetically compounded fertilizers, pesticides, growth regulators and livestock feed additives. To the maximum extent feasible, organic farming systems rely upon crop rotations, crop residues, animal manures, legumes, green manures, off-farm organic waste, mechanical cultivation, mineral bearing rocks and aspects of biological pest control to maintain soil productivity and tilth, to supply plant nutrients and to control insects, weeds and other pests."

Organic fertilizer: By product from the processing of animal or vegetable substances that contain sufficient plant nutrients to be of value as fertilizers.

Precision agriculture: It is the application of modern information technologies and makes use of information system and planning software to provide, process and analyze multi-source data of high spatial and temporal resolution for decision making and operation in the management of crop production.

Radiation: Processes in which energy is sent out as waves and particles through space from atoms and molecules as they rotate, vibrate, and undergo internal change.

Rangeland: Land used for free-grazing livestock.

Re-emitted energy: Radiation of previously absorbed energy.

Reflected energy: Radiant energy that is thrown back from an object, with no change except direction.

Remote sensing: Identifying and observing objects without having any contact with the object. It is feasible through the use of aerial photographs or satellite imageries.

Re-radiation: Radiant emission of energy previously absorbed.

Saline: Containing large concentrations of soluble salts. Operationally defined as the electrical conductivity of a saturation extract of >4 deciSiemens per meter ($dS\ m^{-1}$).

Saline soil: A saline soil contains enough soluble salts so distributed in the soil that they interfere with growth of most crop plants. Ordinarily the saline soil is only slightly alkaline in reaction (pH 7.4 to 8.5) and contains very little absorbed sodium. Saline soils are often recognized by the presence of white salt crusts. As such, a saline soil is sometimes referred to as white alkali soil.

Salinization: The process of accumulation of salts in soil.

Saltation: Particle movement in water or wind where particles skip or bounce along the stream bed or soil surface.

Salt index: It is one of criteria for evaluating the quality of an irrigation water. It can be expressed as follows:

$$SI = (\text{Total Na-24.5}) - [\text{Total Ca} - \text{Ca in } CaCO_3) \times 4.85]$$

Shifting cultivation: A farming system in which land is cleared, the debris burned, and crops grown for 2–3 years. When the farmer moves on to another plot, the land is then left idle for 5–15 years, then the burning and planting process is repeated.

Shortwave radiation: In meteorology, high-energy radiation of the ultraviolet, visible, and near infrared wavelengths and sunshine. Unlike longwave radiation, it passes freely through air.

Soil testing: This refers to chemical tests of the soil that can be made rapidly and with low cost, as compared to conventional methods of soil chemical analysis which are more accurate but more time-consuming and expensive. Soil testing covers both rapid analyses in the field and in the laboratory.

Sustainable agriculture: According to consultative group on international agricultural research (CGIAR), it is the successful management of resources to satisfy the changing needs, while maintaining or enhancing the quality of environment and conserving natural resources.

Universal soil loss equation (USLE): An equation for predicting the average annual soil loss per unit area per year, $A = RKLSPC$, where R is the climatic erosivity factor (rainfall plus runoff), K is the soil erodibility factor, L is the length of slope, S is the percent slope, C is the cropping and management factor, and P is the soil erosion practice factor.

Unsaturated flow: The movement of water in a soil that is not filled to capacity with water.

Urease: A crystalizabale protein enzyme that activates hydrolysis of urea.

$$\underset{\text{Urea}}{NH_2CO.NH_2} + \underset{\text{Water}}{2H_2O} \rightarrow \underset{\text{Amm. carbonate}}{(NH_4)_2\,CO_3}$$

Its richest source is soybean. It is found extensively in jackbeans and numerous fungi. It is marketed as white tablets or powder. It is soluble in dilute alkali. Its isoelectric point is pH 5.5. It is activated by the presence of metals. Urease is found in soils in various degrees. If its concentration is high, urea hydrolysis in soils takes place very rapidly. Urease enzyme taken from soybean or jackbean is also used to determine urea in mixed fertilizers.

Uronite: A sugar with a COOH group.

Virgin soil: A soil that has not been significantly disturbed from its natural environment.

Visible spectrum: Electromagnetic radiations which are within the range of 400–700 nm and visible to the naked eye. The beam formed due to these is called the visible spectrum. It consists of seven colours corresponding to the following wavelength:

Colour of light	Range of wavelength (in nm)
Red	747–700
Orange	585–647
Yellow	575–585
Green	491–575
Blue	424–491
Violet	400–424

X-ray diffraction: An analytical method that involves the exposing of samples to X-rays. The radiation is reflected from the sample in response to structure of the crystals that compose the sample. It is useful in identification of elements in materials and coatings.

Appendices

APPENDIX I

Nutrients, essential for plant growth and forms in which nutrients are taken up by plants

Nutrient	Chemical symbol	Form taken up by plant
Primary Nutrients		
1. Carbon	C	CO_2, HCO_3
2. Hydrogen	H	H_2O
3. Oxygen	O	H_2O, O_2
4. Nitrogen	N	NH_4^+, NO_3^-
5. Phosphorus	P	$H_2PO_4^-$, HPO_4^{-2}
6. Potassium	K	K^+
Secondary Nutrients		
7. Calcium	Ca	Ca^{2+}
8. Magnesium	Mg	Mg^{2+}
9. Sulphur	S	SO_4^{2-}
Micro Nutrients		
10. Iron	Fe	Fe^{2+}, Fe^{2+}, Chelate
11. Zinc	Zn	Zn^{2+}, $Zn(OH)_2$, Chelate
12. Manganese	Mn	Mn^{2+}, Chelate
13. Copper	Cu	Cu^{2+}, Chelate
14. Boron	B	$B(OH)_3$
15. Molybdenum	Mo	MoO_4^-
16. Chlorine	Cl	Cl^-

APPENDIX II

Key to nutrient deficiency symptoms in crops

Nutrient	Colour change in lower leaves
N	Plant light green, older leaves yellow
P	Plants dark green with purple cast, leaves and plants small
K	Yellowing and scorching along the margin of older leaves
Mg	Older leaves have yellow discolouration between veins-finally reddish purple from edge inward
Zn	Pronounced interveinal chlorosis and bronzing of leaves
Nutrient	**Colour change in upper leaves (Terminal bud dies)**
Ca	Delay in emergence of primary leaves, terminal buds deteriorate
B	Leaves near growing point turn yellow, growth buds appear as white or light brown, with dead tissue
Nutrient	**Colour change in upper leaves (Terminal bud remain alive)**
S	Leaves including veins turn pale green to yellow, first appearance in young leaves.
Fe	Leaves yellow to almost white, interveinal chlorosis at leaf tip
Mn	Leaves yellowish gray or reddish, gray with green veins
Cu	Young leaves uniformly pale yellow. May wilt or wither without chlorosis
Mo	Wilting of upper leaves, then chlorosis
Cl	Young leaves wilt and die along margin

APPENDIX III

Average nutrient composition of organic material (oven-dry basis)

Kind of material	Total N	Total P_2O_5	Total K_2O	Total CaO
Compost	1.34	3.30	1.04	0.89
Swine manure	0.81	3.00	0.61	4.75
Carabao manure	0.60	2.05	0.50	–
Cow manure	1.87	2.47	2.11	–
Goat manure	2.81	2.66	1.20	–
Horse manure	3.13	2.80	1.88	–
Sludge	1.87	3.11	0.54	4.30
Vermi-cast	1.86	3.61	1.60	2.21
Azolla	2.76	0.97	2.38	1.09
Rice straw	0.48	0.34	1.58	–

Contd.

Average nutrient composition of organic material (oven-dry basis) *(Contd.)*

Kind of material	Total N	Total P_2O_5	Total K_2O	Total CaO
Coconut coir dust	0.50	0.82	1.26	–
Mud press	2.72	6.20	0.79	–
Distillery slops	0.12	0.25	0.62	–
Garbage ash	0.68	–	1.40	3.45
Water hyacinth ash	0.50	8.06	19.08	–
Factory ash	0.22	2.76	0.94	0.75
Bagasse ash	0.28	0.84	2.00	–

APPENDIX IV

Nitrogen content and C—N ratio of some compostable materials

Materials	Nitrogen content (%)	C—N ratio
Farm residue		
Rice straw	0.3–0.5	80–130
Wheat straw	0.3–0.5	80–130
Barley straw	0.3–0.4	100–120
Maize stalks and leaves	0.8	50–60
Cotton stalks	0.6	70
Sugarcane trash	0.3–0.4	110–120
Lucern residue	2.55	19
Green weeds	2.45	13
Water hyacinth	2.38	17.6
Seaweeds	2.10	–
Azolla	2.5	–
Red clover	1.9	19
Ferns	1.5	25
Flax	1.1	44
Fallen leaves	0.5–1.0	40.80
Grass clippines	2.15	20
Sesbania sp.	2.83	17.9
Neem cake	6.05	4.5

Contd.

Nitrogen content and C—N ratio of some compostable materials *(Contd.)*

Materials	Nitrogen content (%)	C—N ratio
Animal shed waste		
Cow dung	2.0	
Buffalo dung	–	
Horse dung	2.4	
Poultry	5.0	
Sheep	3.75	
Pig	3.75	
Human habitation waste		
Night soil	4.0–6.0	6–10
Urine	15–18	0.8
Digested sludge	5.0–6.0	6
Biogas (ex-cattle) slurry	2.0	20.4
Vegetable residue		
Potato tops	1.6	27
Amaranthus	3.6	11
Cabbage	3.6	12
Lettuce	3.7	–
Onion	2.6	15
Pepper	2.6	15
Tomato	3.3	12
Carrot (whole)	1.6	27
Turnip top	2.3	
Fruit waste	1.5	
Tobacco	3.0	
Forest		
Leaves	0.5–1.0	40–80
Raw sawdust	0.25	208
Rotted sawdust	0.3	128
Mango sawdust	0.3	132

APPENDIX V

Response of pulse crops to inoculation with *Rhizobium* culture under different agro-climatic conditions

S.No.	Crop	Location	Increase in grain yield over control (in %age)
1.	Mung bean	Lam (AP)	14.75
		Dholi (Bihar)	16.49
		Pantnagar (Uttranchal)	4.15
		Delhi	13.33
2.	Urad bean	Pudukottai (TN)	4.21
		Dholi (Bihar)	11.29
		Pantnagar (Uttranchal)	17.21
3.	Cowpea	SK Nagar (Gujarat)	10–36
4.	Pigeonpea	Hissar (Haryana)	5–25
		Pantnagar (Uttranchal)	2–25
		SK Nagar (Gujarat)	9–21
		Sehore (MP)	13–29
		Rahuri (Maharashtra)	3–40
5.	Chickpea	Varanasi (UP)	4–19
		Dholi (Bihar)	25–42
		Delhi	18–28
		Hissar (Haryana)	24–43
		Dohad (Gujarat)	33–67
		Sehore (MP)	20–41
		Badrapur (Maharashtra)	8–12
6.	Lentil	Pantnagar (Uttranchal)	4–26

APPENDIX VI

Key soil and environmental indicators as influenced by agricultural management practices

Soil or environmental indicators	General trend/change	Long-term agricultural practices affecting the indicator
Soil organic matter	Increase	Continuous cropping with well-managed crop residue, zero or minimum tillage, legume-based and other crop rotations, legume plow down (green manure), cover crops, forages
	Decrease	Excessive tillage, summer fallow, crop residue removed or burned
Microbial biomass and biological diversity	Increase or decrease	Same as for soil organic matter
Soil aggregate stability	Increase	Conservation tillage, maintenance of crop residue, forages and legumes in crop rotations
	Decrease	Same as for soil organic matter decrease
Hydraulic conductivity	Increase	Reduced and zero tillage, maintenance of crop residue, forages and legumes in crop rotations-degree and extent of change vary with different practices
	Decrease	Same as for soil organic matter decrease
Soil depth/rooting volume	Increase	Conservation tillage and forage-based crop rotations should reduce erosion and allow soil-forming factors to maintain and rehabilitate top soils.
	Decrease	Excessive tillage, summer fallow cropping system, and crop residue removal or burning are the main agricultural practices that subject soils to serious wind and water erosion resulting in top soil removal
Water quality	Positive or negative	Data are lacking on how soil water quality is affected by different agricultural practices; in general, zero or minimum tillage, forage-based cropping systems, and maintenance of crop residue reduce surface runoff and soil loss to water streams; excessive use of herbicides and fertilizer may result in deterioration of water quality

APPENDIX VII

Proposed minimum data set of soil's physical, chemical and biological indicators for screening the condition, quality and health of soil

Indicators of soil condition	Relationship to soil condition and function; relationale as a priority measurement
Physical	
Texture	Retention and transport of water and chemicals; needed for many process models; estimate of degree of erosion and field variability of soil types
Depth of soil, topsoil and rooting	Estimate of productivity potential and erosion, normalizes landscape and geographic variation
Soil bulk density and infiltration	Indicators of compaction and potential for leaching, productivity and erosivity, density needed to adjust soil analysis to field volume basis
Water-holding capacity (water retention character)	Related to water retention, transport, and erosivity; available water can be calculated from soil bulk density, texture, and soil organic matter
Chemical	
Soil organic matter (total organic C and N)	Defines soil fertility, stability and erosion extent; use in process models and for site normalization
pH	Define biological and chemical activity thresholds; essential to process modeling
Electrical conductivity	Defines plant and microbial activity thresholds, soil structural stability, and infiltration of added water, presently lacking in most process models, can be a practical estimator of soil nitrate and leachable salts
Biological	
Microbial biomass C and N	Microbial catalytic potential and repository for C and N; modeling; early warning of management effects on organic matter
Potentially mineralizable N (anaerobic incubation)	Soil productivity and N-supplying potential; process modeling; surrogate indicator of microbial biomass N
Soil respiration, water content, and temperature	Measure of microbial activity (in some cases plants); process modeling

APPENDIX VIII

Soil order	Area (Million ha)
Inceptisols	134.1
Entisols	78.7
Alfisols	42.2
Aridisols	13.3
Utisols	8.4
Mollisols	1.6
Others	23.7
Total	328.7

APPENDIX IX

Major soil groups of India

Soil group	Area (Million ha)
Alluvial soils	93.1
Red soils	79.7
Black soils	55.1
Desert soils	26.2
Lateritic soils	17.9
Coastal alluvial soils	10.1
Hill soils	3.6
Tarai soils	0.3
Rock out crops	0.6
Others	42.0
Total	328.6

APPENDIX X

Pesticide usage on different crops in India

Crop	Pesticide use (%)	Cropped area (%)
Cotton	54	05
Rice	17	24
Oilseeds	02	10
Vegetables and fruits	13	03
Sugarcane	03	02
Plantation crops	08	02
Other (wheat, coarse cereals, pulses and millets)	03	54
Total	100	100

APPENDIX XI

Technology for production of vermicompost:

- Earthworms ingest vegetable matter, soil etc. and excrete small pellets of finely ground soil (called casts), very rich in nitrogen, phosphorus and potassium. They also turn the soil, thus providing air to microorganisms and the roots of plants. Due to these properties, earthworms can be used to produce vermicompost. The attention has been increasing on breeding of earthworms (vermiculture) and their subsequent use for preparation of vermicompost.
- Typically, a 1.5 hectare farm will need about 12 tonnes of biomass which can be converted into about 8 tonnes of vermicompost which would be sufficient for the given area. Best results are obtained with two eregrine (exotic or foreign) species, *Eisenia foetida* and *Eudrilus eugeniae*. These rapidly convert biomass into casts and also breed very fast. Endemic species (found naturally in India), like *Perionyx excavatus* and *Perionyx sansibaricus* are also suitable but have lower conversion rates and lower breeding rates. Vermicompost units require maintenance of 40%–50% moisture and 20°–30 °C temperature consequently, regular water supply is essential.
- Compared to chemical fertilizers, larger quantities of vermicompost are needed for better soil health and sustained crop production. The yields of crops are usually equivalent to those obtained by using chemical fertilizers. Typically, 1.5 tonnes of feed mixture produces 1 ton of vermicompost in 3 months and continuous cycle has to be maintained for year-round yield of vermicompost. It takes about two or three years of regular use of vermicompost for all benefits to become apparent.
- Pits/troughs can be constructed, either above or below the surface, using brick masonry or stone slabs or even plastic. For a volume of 1 m^3, a pit/trough of 1.6 m length, 1 m width and 0.75 m height is suitable. The number of worms is dependent on the volume: in the above case about 6000 to 7000 worms can be used. If the pits are constructed outdoors, a thatched roof should be built over it as moisture in the 40%–50% range and temperature in the 20°–30 °C range need to be maintained.
- A ditch of standing water around the pit/trough and a wire mesh keeps away predators like insects, cats, dogs and birds. Any organic waste or agricultural residue can be used as feed mixture, e.g. farm yard wastes, green wastes, sugarcane thrash, coir and pith, kitchen wastes. These are mixed with cow-dung in the ratio 8:1 and put in the pit/trough. This feed mixture is allowed to decompose for at least 2–3 weeks. During this time, the mass will heat up. If earthworms are introduced in this period, they will die. Commercially decomposing mixture can be added to facilitate decomposition. Addition of neem powder prevents infestation. 1 kg of earthworms represents 600 to 1000 worms. These can convert 45 kg of wet biomass (40% moisture) in a week's time yielding about 25 kg of vermicompost.

- The partially decomposed biomass in the pit/trough is inoculated with earthworms. Watering is done daily. The worms feed on the biomass, assimilating 5%–10% for their growth and excreting the rest in the form of nutrient rich casts. Once the feed mixture is seen to largely contain casts, it is dumped in a conical heap and left for a few hours. The worms effect at the base and can be easily retrieved for reuse. The rest of the dried material is passed through a 3 mm sieve to collect the casts as vermicompost. A sequential system of vermicomposting has been tried, where various stages are developed in interconnected pits and earthworms migrate from one pit to another.

APPENDIX XII

Technology for production of enriched compost:

Technology for the production of enriched compost has been recommended for field-level adoption through the Ministry of Agriculture and Cooperation. Salient features of the technology are as follows:

- Add 0.5% mineral nitrogen and 5% Mussoorie rock phosphate to the compostable material. Only compostable materials having C—N ratio higher than 50 need addition of mineral nitrogen. Composting of green plant materials or a combination of nitrogen-poor and nitrogen-rich plant materials does not need mineral N input.
- Either spray a suspension of the culture of efficient biodegrading fungal species characterized as microbial inoculants for rapid composting or mix simply 1 part of fresh cattle dung with 2 parts of the compostable material. Fresh cattle dung serves as a highly effective biological inoculant for the composting of organic wastes.
- Bring the compost material to about 70% moisture content and carry out composting in pits or heaps, maintaining appropriate moisture level during the composting period.
- Add inoculum of *Azotobacter chroococum,* if available, to the composting system after 25–30 days of decomposition.
- Provide 4–5 turnings to the composting material at 15 day intervals. A good quality compost rich in humus and plant nutrients will be ready for use in 3–4 months of decomposition.

APPENDIX XIII

Technology for production of phosphocompost:

- Use any organic waste material including crop residues, leaf-litter, partially dried green plant biomass including weeds, animal dung and biodegradable city solid waste. A mixture of various organic materials serves as a good composting material as any single material. Rapidity of composting and the quality of the compost improved if fibrous materials like rice straw and sugar-cane trash are shredded to 5–6 cm size before being put to composting.
- On dry weight basis of the compostable material, fix quantity of Mussoorie rock phosphate either as 12.5% or 25.0% to be incorporated in the compostable organic material. Air-dried organic material may be taken as dry material for calculating the quantity of the rock phosphate to be added for practical purpose. If sufficient quantity of compostable material is available, 12.5% may be chosen as the rock phosphate level. If, however, limited quantity of the organic waste is available, the compostable material should be charged with 25% Mussoorie rock phosphate. Both the levels of rock phosphate will give similar quantity of phosphocompost, only the quantity of phosphocompost to be applied to crops will differ for achieving the same results.
- Use the following formula to determine approximate ratios of the various components of phosphocompost:

Organic waste	:	Animal dung	:	Soil	:	Compost/farm yard manure
8	:	1	:	0.5	:	0.5

- These above components serve as composting mix for the production of phosphocompost. Calculate the rock phosphate to be added, on the basis of whole compost mix, consisting of the organic waste, animal dung, soil and mature compost/farm yard manure.
- The quantity of the compost mix is reduced to 45 to 55% of its original weight under proper conditions of composting. Thus, if 1,000 kg material is composted, one should expect about 500 kg phosphocompost. This finished compost, on an average, contains about 50% moisture and 50% dry compost. This general rule, gained from research experiments, is useful while considering phospho-compost for field applications.
- Make a mixture of the animal dung, soil, farm yard manure and the calculated quantity of rock phosphate, and make a slurry of this mixture in a trough or bucket. Add the slurry to the organic material to be composted and mix the entire mass of the materials as uniformly as possible. Adding rock phosphate to the composting material in the form of slurry ensures very effective transformation of rock phosphate during composting. Maximum transformation of rock phosphate to water-soluble and citric acid soluble forms of P is the hallmark of phosphate.

- Make up moisture level of the compost mix to about 70% and mix the material to ensure that it becomes appropriately moist but not excessively wet. Compost the material for 3 months in a pit or heap by maintaining required moisture level. The composting period can be increased if the compostable material is relatively resistant to microbial decomposition. Give 4 turnings to the material with a rake or spade. First turning should be given 10 days after the material is laid for composting, followed by subsequent turnings at 15 day interval. About 75–80% of the achievable decomposition is accomplished in about 60 days of composting. If composting is carried out by maintaining adequate moisture and by providing recommended intermittent turnings, a good phospho-compost, rich in citric acid soluble P, is ready in 3 months, Appropriateness of moisture level should be checked at each turning.

APPENDIX XIV

Chemical composition of phosphocompost compared with ordinary compost

Component	Ordinary compost	Phospho-compost	Rock phosphate
Total N (%)	1.3	0.95	–
pH	8.6	8.8	8.8
Total P (%)	0.25	3.00	8.4
Organic P (%)	0.05	0.06	0.18
Available P (mg/g)	1.20	0.63	0.18
Water-soluble P (mg/g)	0.31	0.20	–
Citric acid-soluble P (mg/g)	0.52	12.24	2.20

APPENDIX XV

Chemical analysis of organic fertilizer/soil enricher

Chemical properties	Range
Total organic matter (%)	50–65
pH	6.8–7.8
C—N ratio	10:1–20:1
N (%)	1.80–3.0
P (%) total	1.25–2.5
K (%)	0.80–1.50
Ca (%)	1.5–4.0
Mg (%)	0.75–1.0
S (%)	0.40–0.60
Trace elements	All are present
Electrical conductivity	<3.0 dSm^{-1}
Moisture content (%) in organic fertilizer pellets	<10
Moisture content (%) in soil enricher (powder)	18–24

Abbreviations

GHGs: Green house gases
RCTs: Resource conservation technologies
CDM: Clean development mechanism
LUPs: Land use practices
SOM: Soil organic matter
ISFM: Integrated soil fertility management
INM: Integrated nutrient management
SOC: Soil organic carbon
IPNM: Integrated plant nutrient management
BMSF: Biological management of soil fertility
CNG: Compresssed natural gas
FAO: Food and agriculture organisation
SQ: Soil quality
SH: Soil health
ICSWEQ: International conference on soil water environmental quality
UN: United Nations
CEC: Cation exchange capacity
SSSA: Soil science society of America
ESP: Exchangeable sodium percentage
FYM: Farm yard manure
Tg C/yr: Terragram per year
NATP: National agricultural technology project
IPCC: Inter-governmental panel on climate change
PPM: Part per million
GtC: Gigatonnes of carbon

SCS: Soil carbon sequestration
CSA: Clay settling areas
RAPs: Recommended agricultural practices
RMPs: Recommended management practices
INM: Integrated nutrient management
CT: Conventional tillage
NT: No-till
AHAF: Agri-horticulture agroforestry system
SCSPs: Soil carbon sequestration practices
FC: Farming carbon
LUMPs: Land use management practices
UNCCD: United Nations convention to combat desertification
IBI: International biochar initiatives
WTO: World trade organisation
IGP: Indo-Gangetic plains
CA: Conservation agriculture
IPM: Integrated pest management
RDD: Rotary disc drill
PTO: Power takes off
CSSRI: Central soil salinity research institute
UNMDG: United Nations millennium development goals
IFPRI: International food policy research institute
GDP: Gross domestic product
WB: World bank
IMF: International monetary fund
IFAD: International fund for agriculture research
ADB: Asian development bank
PHT: Post harvest technology
RW: Rice-wheat
LLL: Laser land leveling
DWR: Directorate of wheat research
ZT: Zero-tillage
FIRBS: Furrow irrigation raised bed planting system
EFTOs: Eco-friendly tillage options
DSR: Direct seeded rice

LCC: Leaf colour chart
MNCs: Multinational companies
NRAA: National rainfed area authority
STLs: Soil testing labs
STCR: Soil test crop response
DRIS: Diagnosis and recommended integrated system
SSNM: Site-specific nutrient management
STV: Soil test values
IPNS: Integrated nutrient plant system
OFDS: Optimising fertiliser doses
FUEs: Fertiliser use efficiencies
RCM: Research council meeting
CRM: Crop residues management
GM: Green manuring /genetically modified
IWM: Industrial waste material
ECC: Enriched city compost
BNF: Biological nitrogen fixation
BGA: Blue green algae
PSMs: Phosphate solubilising microorganisms
DPL: Dual purpose legumes
LEISA: Low external input sustainable agriculture
GPS: Geographical positioning system
GIS: Geographical information system
VRT: Variable rate technology
NMPs: Nitrogen management practices
LAR: Local ad-hoc recommendations
FP: Farmer practices
PLMZ: Prodcution level management zones
MTI: Miscellaneous topic of interest
ISRO: Indian space research organisation
IARI: Indian agriculture research institute
PCS: Portable personnel computer
GLONASS: Global navigation satellite system
SPS: Standard positioning service
PPS: Precise positioning service

PWM: Precision water management

SSWM: Site specific water management

FCI: Food corporation of India

NABARD: National bank for agricultural rural development

CIAE: Central institute of agriculture engineering

ACPA: Australian center for precision agriculture

SRI: System of rice intensification

IRRI: International rice research institute

ICRISAT: International crop research institute for semi-arid tropics

IISS: Indian institute of soil science

BCR: Benefit cost ratio

TL: Traditional leveling

APEDA: Agriculture processed product export developing agency

IRFT: International research for fair trade

NSOP: National standards of organic products

CCUBGA: Center for conservation and utilisation of blue green algae

NAAS: National academy of agriculture science

BMPs: Best fertiliser management practices

CEC: Cation exchange capacity

NASA: National aeronautics space agency

ISRO: Indian space research organisation

SAC: Satellite application center

CAZRI: Central arid zone research institute

SKUAST-K: Sher-e-Kashmir university of agricultural sciences and technology of Kashmir

IIRS: Indian society of remote sensing

NBSS & LUP: National beauro of soil survey and land use planning

INSA: Indian national science academy

ISCA: Indian science congress association

Index

Afganistan 18
Agro-climatic zone 114
Agroforestry 36
Agro-horticulture 36
Agro-silviculture 36
Alfisol 20
Alkali soils 20, 48
Allahabad 93
Alluvial soil 22
America 9
Arable upland soils 12
Asia 2
Asian development 50
Assam 1
Atmosphere 28

Balanced fertilisation 29, 60
Bangladesh 18
Barmer 1
Barrakpore 20
Bhubaneswar 20
Bihar 1
Bikaner 93
Biochar burial 36
Biogas slurry 3
Biological tillage 42
Biomass 15
Black soils 22
Brazillian cerrado 142

Carbon dioxide 2, 11, 27
Carbon farming 27
Carbon sequestration 3, 5, 12, 31, 45
Cation exchange capacity 8
Cereal-cereal rotation crop residue management 80
Chemical effect 12
Clean development mechanism 3
Climate change 1, 2, 5, 12
Compaction 16
Compressed natural gas 5
Conservation agriculture 41, 42
Conservation tillage 24
Conventional tillage 18
Cotton-sorghum 11
Cover crop 24
Crop rotation 19
Cropping system 76

Decomposition 15
Deforestration 139
Degradation 1
Degraded soils 16
Delhi 114
Desertification 2
Drought 1

Early warning system 1
Earth day 8

Earthquake 73
Ecosystems 2
Energy 48
Environmental friendly technology 9
Erosion 16
European country 2
Evidence 1

Farm yard manure 12
Farming carbon 29
Fiber 8
Food 8
Food security 1, 4, 9
Forestry 10
Future agriculture 25

Genetic engineering 1
Genetic material 1
Germany 46
Global temp 2
Grassland 36
Green manuring 80
Grid sampling 133
Gujarat 157
Gurdaspur 145

Haryana 17
Hayy sceder 47
Himalayan region 139
Himalayas 3
Himachal Pradesh 146
Hindukush 3
Horizon 124
Human health 15
Humans 9

India 18
Indigenous practices 72
Indonesia 72
Industrial waste materials 81
Industrialization 113
Insect 2
Iran 18
Italy 46

Jammu 60

Karnataka 1
Kashmir 1, 2
Kedarnath 1
Kerala 4
Krishi vigyan kendra 156

Ladakh 1
Landslides 76
Large scale 15
Legume in rotation 86
Legume intercropping 85
Legumes 85
Lesar 5
Limiting 15
Living biomass 29

Maize-wheat 11
Management 11
Management practices 62
Management zone 133
Marginal farmers 2
Mellennium development goal 41
Modelling 127
Moisture 124
Mollisols 20
Mountain soils 22
Mulching 17

Nagpur 20
Northern and central Florida 17
Nutrient 48
Nutrient management 63
Nutrient requirement 64

Organic carbon 10
Organic manure 3, 4, 10

Palampur 20
Pantnagar 20

Pearlmillet-potato 31
Pearlmillet-potato-tomato 31
Phosphate rich ore 17
Phosphocomposting 161
Photoscale 122
Photosynthesis 15
Plant biomass 15
Plant residues 70
Pollution 3
Population 113
Portugal 46
Precipitation 3
Precise positioning system 120
Precision agriculture 113
Precision farming 113
Punjab 41

Rainfall 1
Rajasthan 1
Reclaimation 61
Red soils 22
Resource conserving technology 3
Rice-Rice 95

Salinisation 2, 10
Services 10
Shalimar 94
Social security 9
Sodicity 10
Soil biology 157
Soil biota 4
Soil carbon sequestration 15
Soil ecosystem 4
Soil fertility 4
Soil health 7, 9, 18, 60
Soil organic carbon 7, 12, 16, 18
Soil quality 7, 29, 30
Soil restoration 24
South Asia 2
South Asian countries 17
South East Asia 2
Spain 46
Sunflower 31
Sustainability 5
System approach 78

Tamil Nadu 157
Texture 124
Thailand 72
Tobacco 135
Tree 29
Trivendram 19

Ultisol 20
Unexpected rainfall 1
Uttarakhand 1

Variable rate seed 131
Variable rate technology 131
Vertisols 20
Vietnam 72

Water logging 2, 10
Water saving 165
Wild fire 2
Wind erosion 2
Wood 8
World agriculture 10

Zero tillage 3, 29, 48, 52